***f*P**

COREY S. POWELL

THE FREE PRESS NEW YORK LONDON TORONTO SYDNEY SINGAPORE

GOD
IN THE EQUATION

How Einstein Became
the Prophet of the
New Religious Era

LIVERPOOL LIBRARIES & INFORMATION SERV	
LV19708432	
Cypher	03.04.03
291.175POW	£16.99
AL	23/04/03

THE FREE PRESS
A Division of Simon & Schuster, Inc.
1230 Avenue of the Americas
New York, NY 10020

Copyright © 2002 by Corey S. Powell
All rights reserved, including the right of reproduction
in whole or in part in any form.

The Free Press and colophon are trademarks
of Simon & Schuster, Inc.

For information about special discounts for bulk purchases,
please contact Simon & Schuster Special Sales: 1-800-456-6798 or
business@simonandschuster.com

Designed by Chris Welch

Manufactured in the United States of America

10 9 8 7 6 5 4 3 2 1

Library of Congress Cataloging-in-Publication Data is available.
ISBN 0-684-86348-0

CONTENTS

ONE THE GOD OF SCI/RELIGION 1

TWO HOW GOD GOT A JOB IN PHYSICS 15

THREE THE CHURCH OF EINSTEIN IS FOUNDED 47

FOUR THE NEW CARDINALS BICKER IN EUROPE AND AMERICA 81

FIVE EINSTEIN'S PROPHECY FULFILLED 113

SIX THE ERA WHEN THE UNIVERSE CAME FORTH FROM THE HANDS OF THE CREATOR 145

SEVEN HISSES FROM THE MICROWAVES 181

EIGHT THE ANGEL OF DARK ENERGY 209

NINE SALVATION IN THE CHURCH OF EINSTEIN 241

ACKNOWLEDGMENTS 259

BIBLIOGRAPHY 261

INDEX 265

GOD IN THE EQUATION

LIVERPOOL CITY COUNCIL

CHAPTER ONE

THE GOD OF SCI/RELIGION

THE WORLD'S NEWEST spiritual center is a long way from Mecca or Jerusalem, Vatican City or Lhasa. It lies at the remote summit of Mauna Kea, a million-year-old mountain of lava and ash jutting nearly three miles above the tropical Hawaiian shore. While the old sites still hold a sacred place in the imaginations of countless billions, their spiritual wisdom hasn't changed in centuries. Hearing the latest gospel requires a pilgrimage to the top of this hulking, dormant volcano.

Antiseptic white domed buildings, mysteriously unmarked, interrupt the desolate landscape. There are no spires, no stained glass, no columns or gilded doorways to welcome the visitor. Inside, the high priests of science and their electronic surrogates peer through the dry, rarefied atmosphere into the vastness above. A trek to this remote pinnacle of astronomical power begins on the lightly traveled Saddle Road, which runs between Mauna Kea and its active twin, Mauna Loa, on Hawaii's Big Island. Your rental car is wheezing by the time you reach the stopping point halfway up, at Hale Pohaku. Even after a night of acclimation, your own lungs are working double time when you reach the top, an altitude of just under fourteen thousand feet. At night the stars appear strangely dim, because your retinas are starved for air and unable to pull their normal duty. The sky swims with the blackness of a near faint, similar to the kind of negative rush you see when you stand up too quickly. Take a deep breath of pure oxygen from a tank and all the visual chemistry falls back into place. Then, a revelation: The sky blooms with light, the universe made manifest by a dose of rudimentary medical technology.

The two Keck telescopes are the supreme oracles of Mauna Kea. After a half hour in the dark, a healthy person's pupils open not quite one-third of an inch, and behind them the retinas store about one-tenth of a second of visual information. Keck I and its newer twin, Keck II, maintain an unblinking gaze thirty-three feet across and can hold it for hours. Their thirty-six-piece, segmented mirrors gather billion-year-old light from faraway quasars and galaxies, amassing the raw information to answer questions otherwise unknowable to mere mortals. Photon by photon, the Kecks are validating the new way of understanding the world.

The word coming down from Mauna Kea is not traditional science. It is too grand in scope, embracing all of space out to the edge of the universe and all of time back to the moment of cosmic origin. It is empirical, but it knowingly overreaches, describing particles that have never been detected, fields that have never been felt, and regions of space that have never been seen. It utterly dwarfs human conceptions, much like an omnipotent deity. Yet this celestial form of enlightenment also bears little resemblance to Judaism, Christianity, Islam, or any other old-time religion. In place of the unity of God, it seeks out simplicity of explanation. In place of a central dogma, it rests on the falsification of theory through empirical data. It develops its own entrenched doctrines, but it also provides the tools with which to discard them. This new faith has acquired millions of converts and permeated every corner of American culture. It has changed our world, but until now it hasn't had a name.

Call it sci/religion, because it blends elements of the experimental and the mystical. The name also works as a pun on two defining aspects of modern science. In quantum theory, the Greek letter *psi* represents the fundamental uncertainty of measurement. At any moment, a subatomic particle does not have a single, well-defined position; instead, it has a statistical blur of potential positions. In essence, the uncertainty principle means a particle can be in two places at the same time, allowing interactions that would be forbidden according to classical physics or common sense. This quantum rule bending permits the nuclear reactions that cause stars to shine, and in some current cosmological models it even explains the origin of the universe. So: "psi religion." Sci/religion also evokes scientists' wariness about openly discussing the metaphysical strains that are increasingly obvious in their work. They shy away from questions about their personal faith, fearing that any answer will only make them look foolish or reactionary. When you ask them if they believe in God, they always respond the same way, with a sigh. So: "sigh, religion."

The founder and greatest prophet of sci/religion had no such qualms about finding common ground between the material and the mystical. Albert Einstein recognized the search for truth as an inherently spiritual endeavor. "Everyone who is seriously involved in the pursuit of science becomes convinced that a spirit is manifest in the laws of the universe—a spirit vastly superior to that of man," he explained to one of his students in 1936. Explicitly and implicitly in his work, Einstein preached the doctrines of unity, simplicity, and universality. These principles are the guiding lights of sci/religion. Few of his followers speak as openly as he did, but their actions give them away. Just look at the beliefs that motivate their experiments, their equations, and their journal articles. Look at their research on Mauna Kea. They worship in the Church of Einstein.

Like many people of my generation, I grew up immersed in the faith of sci/religion. As a child I marveled at the drawings of swirling nebulae and colliding galaxies in classic books such as *The World We Live In.* Later I read about quasars in *Astronomy* magazine and puzzled over descriptions of curved space-time in *Scientific American.* Imagining the infernal fireball of the big bang sent chills down my spine back then, and it still does so today. Later, digging into the history of science, I learned how these ideas have kept changing as one theory failed a crucial test and a new theory came along. That realization only strengthened my appreciation for the mystical power of science. Each blip of starlight arrives loaded with meaning. The priests of sci/religion steer their telescopes, observe, analyze, and draw up new theories. They know they will never attain full understanding but labor away, confident that they will arrive ever closer toward cosmic enlightenment. That endless pilgrimage gives purpose to life.

Four years ago, I heard that one of Einstein's disciples had experienced a breakthrough in the sci/religious faith. Saul Perlmutter, a hard-driven and deceptively boyish cosmologist at California's Lawrence Berkeley Laboratory (LBL), had been approaching the Keck Observatory with new versions of some of our oldest queries. Will the universe continue forever, or will it someday come to an end? Do the heavens operate according to the same physical rules as the terrestrial realm? Above all, is there more to the universe than meets the eye? Three thousand miles away, in his spare hilltop office above the San Francisco Bay, Perlmutter pored over his precious data. The answers he sought might already be encoded somewhere in the buckets of starlight gathered by the Kecks.

Perlmutter approached the great telescopes armed with a cunning plan to force those secrets into the open. In the late 1980s he developed a strategy to measure how the expansion of the universe is slowing down—one of the fundamental pieces of information sought by modern cosmologists—by studying the light from distant exploding stars. Many people had proposed this approach, but

Perlmutter was the first to develop the analytical techniques and computerized tools that could translate the brief flaring of those stellar detonations into meaningful messages. The plan worked beyond his wildest imaginings. By early 1998 he had collected enough observations to see signs of wonder.

The universe is not slowing down under the pull of gravity, as astronomers had naively assumed. It appears to be accelerating, galaxies rushing apart faster and faster under a mysterious repulsive influence. Perlmutter's collaborators were understandably skeptical, but they could find no flaw in his work. Neither could his scientific competitors, led by the similarly youthful Brian Schmidt of the Research School of Astronomy and Astrophysics at the Australian National University. Schmidt's team got off to a later start but pursued a similar line of attack and arrived at essentially identical results. By the summer of 2000, two balloon-borne instruments called BOOMERANG and MAXIMA added to the stack of supporting evidence in favor of a runaway, accelerating universe.

The oracles had spoken. Haltingly, Perlmutter and the other researchers accepted the message. The visible universe is only one small part of what is out there. Even the invisible material, the eerie "dark matter" that knits together clusters of galaxies, is a secondary element. Every known form of matter produces an attractive gravitational force over large scales. Cosmic acceleration indicates the universe must also contain something that produces a strong repulsion. Albert Einstein considered such a mystery component, which he called "the cosmological constant" and denoted by the Greek letter lambda but later dismissed it as strange and unproven. Perlmutter's findings implied Lambda is real. It is so powerful that it overwhelms the inward pull of all the galaxies and dominates the universe. At face value, this discovery was the astronomical equivalent of learning that the United States is run not by the president and Congress, but by an unsuspected race of elves who hide inside tree stumps. But the reaction to the news was nearly as significant as the result itself. When Perlmutter announced his results, nobody seemed terribly shocked. His colleagues received the word from Mauna Kea calmly and warmly. Soon they had given this antigravity force a nickname, "dark energy," and added it to their regular vocabulary. As I watched this response, I fully appreciated for the first time how thoroughly cosmologists have embraced the faith of sci/religion.

The observers, the folks who spend agonizing nights scrutinizing the faintest flecks in the sky, nodded in encouragement. They agreed that the accelerating universe was a provocative discovery, gently cautioning that measuring the rate of cosmic expansion is difficult research, prone to many possible errors. They had seen plenty of unexpected phenomena before and were always prepared to be caught off guard again. Meanwhile, the theoretical cosmologists—the mathematical thinkers who spin physical tales about the origin and fate of the universe—responded with an equal mix of enthusiasm and sangfroid. Not only had they already considered the possibility of Lambda, they had gone a step further and assumed something like it had to exist, because Lambda provided crucial symmetry to theories that otherwise appeared out of balance. Even the public took the reports in stride. Writers in the newspapers and popular magazines trumpeted Perlmutter's findings as startling and bizarre, but the coverage quickly returned to a familiar tone of reverent wonder.

There were plenty of follow-up news stories and articles, of course. *Science* magazine, the leading American general-science journal, touted the finding as its "Breakthrough of the Year" at the end of 1998. "Scientists and philosophers will be grappling with the implications for years to come," the magazine promised in the standard grandiose-yet-deadpan language of academic salesmanship. Cosmologists developed competing theories about the energy behind the runaway expansion. Observers suggested new ways to make sure the supernova results were not flawed. What was missing was a visceral reaction—amazement, confusion, perhaps even outrage—that the universe had pulled a fast one on us. Earlier trailblazing discoveries, such as Einstein's theory of relativity and the discovery of the expanding universe, engendered fierce scientific debate that spilled over into the public consciousness. Why the great calm this time around?

The simple answer is that science had transformed into sci/religion. Those earlier discoveries were so wrenching precisely because they were the ones that effected the change. Quantum physics introduced the idea that space is never really empty but seethes with potential energy and matter. Most important of all, the general theory of relativity smashed the false idols of classical science. Time, dimension, and mass are not fixed entities, Einstein declared, and must be replaced by new concepts that conform to a deeper reality. Sci/religion unfolded from this prophecy. In the gospel according to Einstein, space can bend and stretch. These improbable ideas soon found validation in the discovery of the expanding universe. In his vision of cosmic unity, Einstein linked all of space through his equations of general relativity and connected every mass with every other mass. A dandelion seed floating over a suburban lawn influences a distant quasar, and vice versa. To make this vision into a coherent picture of the universe, Einstein created the hypothetical propulsive component, Lambda, and folded it into his omnipotent equations.

Scientists accept the nonintuitive notions of modern physics because they match up with experimental data. But scientists embrace these ideas emotionally because they promise a transcendent understanding of the true nature of the universe. Call it prayer in the Church of Einstein. When Perlmutter and Schmidt uncovered evidence of cosmic antigravity, the discovery did not contradict scientific expectation. It affirmed what scientists already believed and what they already felt. Cosmologists had been expecting, even yearning for, something like Lambda to fill in the gaps in their models. Lambda deepened the blissful sensation that sci/religion has transcended the human world.

As the apotheosis of sci/religion, Lambda perfectly illustrates how close science and religion have been all along. In the old-time religions, there is more to the world than matter. There is heaven and hell, there is the immortal soul, and above all there is the unseen and unknowable divine Creator. In science, there has been a parallel search for immaterial forces that animate the world. Religion searches for knowledge about the intangibles through the reading of Scripture. Science carries out its pursuit through the reading of experimental evidence. Both assume that such readings will lead ever closer to an ultimate, but perhaps never fully attainable, truth. In both worlds, the drunken excitement of enlightenment is fundamentally the same.

Claiming parallels between science and religion tends to offend people on both sides of the fence. Scientists disapprove of the implication that their work is guided by dogma rather than data. Theologians fear that attempts to link religion to the empirical study of the world undermines faith. After the church's battle with Galileo in the seventeenth century, the two sides worked out a rough line of demarcation: Science would tackle the material world, while religion would take responsibility for matters moral and spiritual. Saint Augustine of Hippo and Saint Thomas Aquinas had established the basic argument that the Bible was not intended as a textbook on the physical workings of the world. In essence, the church accepted Galileo's argument—voiced originally by a member of the clergy, Cardinal Baronius—that Scripture explains "how one goes to heaven, not how the heavens go." Yet violations of the borders still occur, mostly carried out by adherents of the old-time religions seeking to defend their turf.

There is, first of all, the drearily familiar battle between biblical literalists and schools that teach evolution. (Big bang cosmology and the geological history of the Earth also contradict Genesis, but they are not as commonly taught, nor are they as emotionally charged as human origins.) This disagreement led to the Scopes trial in 1925 and the 1999 Kansas School Board decision—since reversed—to strike evolution from the state's science curriculum. In essence, the creationists assert that science has overstepped its boundaries by proposing theories of origin for humans and for the universe, and they seek to reclaim the material world in order to prevent any conflicts with the Bible. The most extreme creationists hold that the Earth is six thousand years old and reject any evidence that could interfere with that belief. In addition to the obvious rebuttals from the fossil record, these creationist arguments disintegrate on their own logic. If you take every word of Scripture at face value, then you have ridiculous situations such as Noah trying to cram thirty million species into his ark. On the other hand, if you accept that some of the Bible is allegorical or metaphorical, why try to make any of it function as a scientific textbook?

Then there is the reverse argument, that modern cosmology actually proves the story of Genesis and, by extension, the existence of God. In his famous book, *God and the Astronomers,* former NASA director Robert Jastrow helped promote this idea with his much misinterpreted quote, "For the scientist who has lived by his faith in the power of reason, the story ends like a bad dream. He has scaled the mountain of ignorance; he is about to conquer the highest peak; as he pulls himself over the final rock, he is greeted by a band of theologians who have been sitting there for centuries." Jastrow, a self-professed agnostic, well understood that the similarities between the Bible and the big bang are mostly superficial. Any description of cosmic creation will have a beginning or it will not. Either way, the scientific version would correspond loosely to one of the world's major religions. And as many cosmologists point out, the discovery that the universe evolved from a hot, dense state does not prove a divine agent was responsible for establishing those initial conditions.

Many creationists look past the big bang and make the more subtle argument from design: The cosmic laws are so intricate and perfectly tuned to allow the existence of intelligent life that they must be the handiwork of a divine being—an argument also commonly applied to the biological world. The enormous hole in this reasoning is that it depends on a very human, subjective judgment regarding which aspects of the universe are so wonderful that they could only have come directly from God. But as human knowledge progresses, the boundaries change constantly. The circular motion of the heavens no longer seems miraculous once you understand that the Earth rotates. The jagged thrust of the Himalayas seems quite natural once you recognize that continents move and collide. Knowing about DNA instantly takes the mystery out of heredity and mutations. The argument from design is a modern variant of the old ontological argument, which states that God must exist because something had to put the concept of God into our heads. In its contemporary form—that God must exist because nature seems so incomprehensibly wonderful to us—the reasoning is equally unsatisfying.

These attacks spring from a misguided premise. They assume science has no place for faith, so religion must create one. But sci/religion abounds with faith. It doesn't simply reduce the world to ordinary, material explanations, as many critics of science contend. Sci/religion constantly carves out new space for the extraordinary and the intangible as it proceeds in its relentless search for underlying reality. Its mystical visions are as fantastic as anything in the Bible, but they are fundamentally different. The modern sci/religious liturgy has tremendous credibility because it rests on the same principles—testable theories and repeatable observations—that have produced so many other tangible scientific and technological advances. That is why more people today probably believe in black holes than believe that Moses literally parted the Red Sea to lead the Israelites out of Egypt.

At the same time, old-time religion is losing its grip on its home turf, the realm of ethics and morals. Traditional religious belief is growing increasingly marginal in an age when people look to everything from angels to Oprah to psychotherapists for guidance. Even among many people who maintain the appearance of observance, democracy and capitalism have eaten away at religion's uniform moral authority. Dietary rules fade away; patriarchal practices get watered down or discarded. These days, presidents look to professional ethicists for guidance on questions of biotechnology issues, issues that themselves seem to infringe on God's old creative domain. Creationists don't hate evolution per se. They hate the implied loss of authority of Christian values. The fact that creationists use scientific evidence to support their cause shows how far the balance has shifted. Try to imagine scientists feeling compelled to bolster their position by insisting that schools cite scriptural passages supporting the idea that the universe is governed by empirically knowable laws.

In fact, religion has been in retreat for centuries, both before and after the line of demarcation. Theologians repeatedly set out to define and defend their faith by clarifying the distinction between the earthly and the divine, but in the process they left more and more room for the empirical study of the world. Saint Augustine argued that using the human senses to study nature is a valid way to explore the glory of God. The twelfth-century Jewish philosopher Moses Maimonides sought a spiritual system that was compatible with Aristotle's model of the universe, "as a means of removing some of the doubts concerning anything taught in Scripture." In the thirteenth century, Saint Thomas Aquinas folded Aristotelian physics into Christian belief, showing that the two could peacefully coexist. Baruch Spinoza, the seventeenth-century theologian, introduced the radical concept of a God that does not interfere with the operation of the world but is fully defined by His laws of nature.

Sci/religion has advanced to fill the void by offering its own forms of ecstasy. The most intense feelings arise when sci/religion reaches highest and farthest, striving to grasp the most remote workings of the cosmos. Aristotle proposed a fifth element to explain what keeps the sun, moon, and planets circling the Earth on their ceaseless circuit. He assumed that the heavenly element was something perfect and divine, hence removed from our flawed world. Twenty centuries later, Isaac Newton made a huge advance toward bringing the heavens within reach. He explained that all matter has an intrinsic property, inertia, which causes a moving object to keep moving. He recognized gravity as the universal attraction that controls everything from falling apples to the orbiting moon. He tethered us to heaven but still imagined that God was hidden—not in rotating spheres but in the fundamental, unmoving structure of space.

Newton's theory of gravity provided a mathematical description of how the attraction works but did not explain what gravity is. The gravitational field seemed almost magical, spreading its influence through empty vacuum. Rival scientists, and Newton himself, expressed philosophical reservations about this process of "action at a distance." More troubling, gravity seemed to be too powerful. If everything pulls on everything else, Newton wondered, what holds the universe up? He tentatively solved the problem by assuming the universe is infinite. This was both a theological and a scientific fix. An unbounded universe could not collapse toward some central point under the spell of gravity, he believed, and the endless expanse of the stars reflected the infinite glory of his God.

Einstein took these ideas a crucial step further. In his general theory of relativity he made space an active partner with matter, giving the intangible equal status with the tangible. Matter curves space-time, and that curvature is what we feel as gravity. One kind of spookiness went away, only to be replaced by another. When he expanded these ideas to cosmic scale, Einstein became convinced that the inertia of every object is linked to the curvature of the entire universe. That linkage made sense only if the universe were finite; otherwise there would be no specific spatial background against which to measure the progress of an apple from its branch to its resting place on the ground. To explain gravity and inertia, Einstein erased Newton's infinity. But all of a sudden, gravity was again out of balance. The intangible wanted to take control, making the universe collapse in on itself. This is why Einstein invented Lambda: to tame the spiritual forces and keep the sky from falling.

In Einstein's finite universe, there is no escaping the authority of science. There is no heaven where miracles can occur, no infinite space to harbor Newton's God. The old-time religions proposed that prayer and ritual observance create a link between the individual and a willful deity. Einstein presented the possibility of a cosmic connection based on an intellectual comprehension of the rules of reality. To him, these rules and God were one and the same. His gospel of sci/religion led him to the same point where Spinoza had made his religious last stand. "I believe in Spinoza's God who reveals himself in the harmony of all that exists, but not in a God who concerns himself with the fate and actions of human beings," Einstein said. Lambda represented his search for a harmonious God in his equations.

Einstein eventually renounced Lambda, but never doubted his faith in a mathematically beautiful, comprehensible universe. Lambda, meanwhile, has resurfaced again and again because of its spiritual power. It bestows exquisite balance onto today's cosmological models and so demonstrates the mystical power of sci/religion: its ability to explain the entire universe in a tidy set of mathematical concepts. Most of those touched by Lambda have probably never even heard of it. The total number of people who understand all the details of modern cosmology is quite small, after all. But the number of people who accept and follow Einstein's gospel is huge. The same empirical methods that conquered the most remote galaxies have also led to electric toasters, computers, and nylon panty hose. Einstein's cosmic religious sense, a feeling of "the nobility and marvelous order which are revealed in nature and in the world of thought," has triumphed in every aspect of our lives, from the mundane to the sublime. Sci/religion is no longer only about how the heavens go. It is also about our relationship to the heavens.

Lambda expresses the Word from the great white domes on Mauna Kea. It encapsulates cosmologists' wildly optimistic belief that the universe is knowable and that we are right now on the verge of an all-encompassing understanding. Lambda's ethereal nature is integral to its inspirational appeal. The story of Lambda is the story of the secret faith that keeps sci/religion, and the human spirit, pushing ever onward.

CHAPTER TWO

HOW GOD GOT A JOB IN PHYSICS

THE GREAT PROPHET of sci/religion has an intimidating, otherworldly image. Albert Einstein's name conjures up unkempt shocks of gravity-defying grey hair framing a thought-lined brow; time travel, black holes, and other science fiction–tinged exotica; perhaps the cryptic equation $E=mc^2$ and its explosive realization in the atomic bomb. Look more closely, however, and Einstein transforms back into a normal man driven by familiar impulses. He wanted to know where our world comes from and why it works the way it does. He wanted to understand how the remote stillness of the heavens relates to the erratic, ever-changing events here on the Earth. Above all, he wanted to know if the answers to these questions would bring him closer to God.

Einstein was far from the first to head down this path. Greek philosophers had wrestled with many of the same questions two dozen centuries earlier. They looked past the popular mythologies of the day, which attributed the inexplicable vicissitudes of weather, crops, and disease to a cantankerous community of gods. Led by the libidinous Zeus and his volatile wife, Hera—who happened to be his sister as well—these immortals ruled the world in accord with their ever-shifting moods. Mount Olympus was geographically close to the mortal realm, and the gods who lived there were likewise nearly human save for their extraordinary powers. These characters made for entertaining storytelling, but their behavior merely reflected the chaos of everyday life, it did not explain it.

More meaningful answers seemed to lie elsewhere. Like Einstein, the Greeks sought truth in the purity of mathematics and in the heavens, where an entirely different kind of order prevailed. When the sun set, the sky glowed with innumerable pinpoints of light. It was impossible to ignore the majestic enigma of this other realm in an age when fire was the lone controllable source of nighttime illumination. The array of stars remained fixed within the vault of the heavens, unchanging from generation to generation, while the whole bowl-like firmament completed one perfect circuit each year and one somersault each day. The sun and moon wheeled their way across the star-flecked background. And a few rule breakers—the planets, literally "wanderers"—strayed across the skies in maddeningly complicated ways. Many ancient cultures sought the hidden patterns in these motions, but the Greek natural philosophers developed the uniquely creative solutions that still inspire and influence modern cosmology.

Eudoxus of Cnidos, the father of Greek mathematical astronomy, made one of the first systematic attempts to account for the irregular regularity of the heavens. His treatise, *On Velocities,* is long lost, but its ideas survive through his influential followers. Nearly four centuries before Jesus' birth, Eudoxus proposed that the universe consists of a series of nested, transparent shells, which carry the sun, moon, and planets as they rotate. The fixed stars sit on the outermost sphere. The Earth sat at the center of it all, but the spherical shells defined the heavens as distinct from the Earth's murky, tumultuous elements. This spherical cosmology existed to perform what historians of science call "saving the appearances"—

that is, simulating the appearance of the natural world rather than explaining how it operates. But in this early era, Eudoxus already adhered to the principles of unity and simplicity that would eventually guide sci/religion to dominance.

In its basic elements, Eudoxus' model of the universe took inspiration both from Pythagoras, who had declared the sphere the perfect shape, and from Plato, who had come to embrace the idea that each component of the heavens resides on its own concentric sphere. Eudoxus reputedly bristled at the highly abstract nature of Plato's teachings—he ended up founding a rival school based on a more rigorously observational approach—but adapted these ideas to his own ends. He added additional spheres and coupled their motions, so that the whole system could explain in detail how the various bodies roamed among the stars. The resulting system was both aesthetically appealing and amenable to geometric analysis.

This was not intended as a realistic, physical model of the universe. Most likely Eudoxus regarded the spheres as useful concepts rather than real objects. It was not particularly brilliant as a predictive model, either: it required a total of twenty-seven independently moving shells, and still its inaccuracies would have been immediately evident to the naked-eye observers of the day. What Eudoxus crafted was a descriptive system that accounted for the general nature of the solar, lunar, and planetary motions. As such, it marked a small but incredibly important step toward the creation of a kind of cosmic religion, one that aspired to truth by blending philosophy and spiritualism with mathematics and observation. In Eudoxus, ancient beliefs mingle with a recognizably modern hunger for scientific certainty. "Willingly would I burn to death like Phaeton, were this the price for reaching the sun and learning its shape, its size, and its substance," he swore.

Judged purely on how long it survived, Eudoxus' system of spheres would have to be considered the most successful cosmology in the history of thought. It was not fully overthrown until Johannes Kepler discovered that planets move in elliptical paths, not circles, and Galileo Galilei established observational evidence that those paths go around the sun, not around the Earth. Both of those epiphanies occurred during the first decade of the seventeenth century, roughly two thousand years after Eudoxus. Our modern big bang interpretation of cosmic history is an infant by comparison, scarcely fifty years old. Although the spheres of Eudoxus emerged from a pagan, mathematical tradition, Saint Thomas Aquinas absorbed them into church doctrine. By the time scientific knowledge finally destroyed that Earth-centered cosmology, it was so entrenched in Christian thought that the transition triggered an international religious crisis. The loss of those spheres destroyed the notion of the heavens as a physical place whose divinity increased with its distance from this flawed world, and it cast God out of His home in the firmament. It took Einstein and his second cosmological revolution to revive the old spiritual feeling that a single, monotonic line of existence runs from the Earth to the farthest star.

During their long, wildly successful tenure, the spheres of Eudoxus underwent numerous refinements, translations, and adaptations. His disciple Callippus of Cyzicus touched up some of the observational failings of the model by adding six additional spheres. But by far the most famous elaboration was carried out by one of the greatest of the Greek philosophers, Aristotle. As a student, Aristotle had studied at Plato's academy while Eudoxus was running it. Eudoxus' ideas clearly landed on fertile ground, but Aristotle added many innovations of his own. One of the hallmarks of Aristotle's philosophy was emphasis on the importance of empirical data—when it suited him, at least. Observation often seemed to serve as a justification more than as inspiration for his ideas. Bertrand Russell once quipped, "Aristotle maintained that women have fewer teeth than men; although he was twice married, it never occurred to him to verify this statement by examining his wives' mouths." Still, Aristotle paid much more attention to the world of the senses than did Plato or Eudoxus, which led him to devise an even more complex model of the cosmos. To reconcile the general motions of the planets with a mechanically plausible, dynamic cosmos ruled by circular motion, Aristotle described a set of fifty-five crystalline spheres, rather more like a set of uncoordinated wedding gifts than the sort of economical structures one might associate with celestial harmony.

In his great cosmological work *On the Heavens (De Caelo)*, Aristotle developed a theory that would not only describe the motions in this crowded sky, but also explain them. He postulated that the heavenly realm is composed of a fifth element, called "ether." Unlike the four elements of the human world—earth, fire, water, and air—ether naturally follows circular motion. Thus there is no problem of inertia for the cosmic spheres because the heavens, once set in motion, never run down. Ether was the essential intangible element that allowed the spheres to maintain their form and the motions. It was fundamentally unlike later scientific intangibles because it was fully cut off from the human world. Nothing like ether existed in our realm; ether functioned only on the contingency of the Creator who gave it an initial impetus. Aristotle envisioned the outermost shell of heaven, the one containing the "fixed" stars, as the edge of everything. The whole must be finite, he argued, because it clearly moved in a circle about the Earth, and an infinite sphere could not complete its rotation in a finite amount of time: "It is impossible that the infinite should move at all." He considered the whole system eternal, in pointed contrast with earthly affairs marked by an endless series of cycles of renewal and decay. "Whatever is divine, whatever is primary and supreme, is necessarily unchangeable," he wrote.

By transforming Eudoxus' spheres into physical entities, Aristotle made an explicit connection between mathematics and God. The perfect simplicity of spherical motion both expresses and defines the divinity of the celestial world. Decoding that motion therefore becomes a stairway to heaven. As Aristotle wrote in his *Metaphysica,* "The mathematical sciences particularly exhibit order, symmetry, and limitation; and these are the greatest forms of the beautiful." This exaltation of order and symmetry lives on in modern science, most visibly in the mystical extremes of particle physics and cosmology.

The Aristotelian universe, with its clear distinction between earthly and heavenly elements, established an extremely influential picture of a hierarchical universe. As he declared in *On the Heavens,* "We may infer with confidence that there is something beyond the bodies that are about us on this earth, different and separate from them; and that the superior glory of its nature is proportionate to its distance from this world of ours." In this scheme, the Earth is at rest; the planets can move, but in a flawed multiplicity of motions; only the "first heaven" attains the full perfection of the circle. The Catholic Church, under the philosophical guidance of Aquinas in the thirteenth century, fused this cosmological model with Christian theology, so that the celestial spheres became the literal abode of the angels. This vivid picture is still a staple of popular culture, but it has cost the church dearly—once when Copernicus and Galileo argued powerfully that the Earth is not at the center of the universe, and again when Einstein developed the first comprehensive cosmological model rooted in physics. From that point on, Aristotle's "superior glory of nature" belonged to science, not to religion.

The mechanical aspects of Aristotle's dynamic, bounded universe also proved remarkably durable. Ether calls to mind its modern counterpart, the dark energy or Lambda often invoked by cosmologists. Like ether, Lambda is an unfamiliar element whose invisible influence supposedly controls the observed heavenly motions. Einstein wanted to design a universe that was static and unchanging, an echo of the Aristotelian dictum that "that which is divine must be eternal." But Lambda exists within the sci/religious doctrine of falsification through observation—and indeed, observation nearly killed off Lambda a decade after Einstein invoked it. Aristotle, on the other hand, was largely free to go wherever his intuition carried him. The commonsense appeal of Aristotle's cosmology also explains its longevity. His description of the spherical shape of the universe still has popular resonance because it seems so natural. Look up at the night sky and it *looks* like a dome of stars. No wonder you can still walk into a science hobbyists' store and buy a model of the sky showing the stars embedded on a spinning transparent globe. Memories of the Aristotelian spheres repeatedly surfaced in science as well. Einstein's first attempt to derive a cosmological model built around his general theory of relativity yielded a universe having a very familiar finite but unbounded shape: a sphere.

If the sphere of stars defines the outer limits of existence, as Aristotle argued, then it must have a definite extent. Thus Greek cosmology initiated a new way to evaluate the extent of God's glory, calculating the size of the universe. The answer depends on the number of spheres and on how tightly the various spheres could be packed together. Efficient packing became a lot more difficult when Claudius Ptolemaeus, better known as Ptolemy, attached small secondary circular motions onto the large circles marked out by the planets as they moved across the sky. These "epicycles" required considerable maneuvering room between spheres. Epicycles further compromised the geometric elegance of Eudoxus' system, but they did do a much better job at predicting planetary motions. The Islamic astronomer al-Farghani, writing in the ninth century C.E., estimated the minimum possible size for the solar system assuming all the parts were arranged as compactly as possible. By his reasoning, Saturn lies at a distance of about twenty thousand Earth radii, or about eighty million miles. He was off by just a factor of ten, which has to count as a major coincidence considering he was working from an entirely erroneous model.

But what if the Earth circles the sun rather than the other way around? The Pythagoreans had proposed that fire, not earth, should lie at the center. Although Aristotle had derided the idea, Aristarchus of Samos went further and proposed a sun-centered universe around 280 B.C.E. This heliocentric, or sun-centered, model did not attract much of a following at the time, as it seemed to contradict both logic and common sense. Even the most casual sky gazer knows that the stars appear immobile relative to one another throughout the year. As long as the Earth is motionless, there is no problem. In fact, the unchanging appearance of the stars was one of Aristotle's key arguments that the Earth does not move. If the Earth swings around the sun, however, its change in position must slightly distort the constellations and make some stars appear to grow brighter and dimmer over the course of the year. No such variations are seen, so either the heliocentric model is wrong or else the distances to the stars are enormous compared with the size of the Earth's orbit around the sun. It is hard to fault the ancient Greeks for failing to appreciate the vast gulf between the planets and the starry firmament. Astronomers now know that the nearest star, a faint red sun named Proxima Centauri, lies some 267,000 times farther away than the sun; its apparent back-and-forth motion is 300 times too small to be seen by the naked eye. The incredible emptiness of empty space is why perspective effects are completely invisible.

So the Aristotelian system reigned supreme in the Western world, refined by Islamic astronomers and reiterated in the teachings of medieval universities, until it faced an unlikely challenger, a Polish scholar named Nicolaus Copernicus. Around 1510, Copernicus resurrected the sun-centered universe in an unpublished work, *Commentariolus,* which he shared with a few friends. He developed these ideas at much greater length in his book *De Revolu-*

tionibus, published shortly before his death in 1543. He had delayed making his views publicly known, fearing that it would arouse the wrath of the Catholic Church: after all, Joshua commanded the sun, not the Earth, to stop moving during his battle at Jericho. But Copernicus was no iconoclast, at least not an intentional one.

As a sign of faith, Copernicus dedicated his book to Pope Paul III and preemptively defended his ideas against rabble-rousers who would quote the Bible to attack ideas that, he believed, lay outside the bounds of religion. "If there should chance to be any mathematicians who, ignorant in mathematics yet pretending to skill in that science, should dare, upon the authority of some passage of Scripture wrested to their purpose, to condemn and censure my hypothesis, I value them not, and scorn their inconsiderate judgment," he wrote. And, like his predecessors, he bowed before the majesty of the sphere and believed Plato's dictum that the planets followed uniform circular motion. But he wasn't satisfied with the philosophy or the aesthetics of the geocentric arrangement of the heavens. By placing the sun in the middle, Copernicus eliminated some of the unappealing epicycles from Ptolemy's cosmic model. He also created a more unified scheme by giving Mercury and Venus full orbits around the sun, granting them equal status with the other planets. In an Earth-centered model, these two are arbitrarily constrained in their motions to explain why they always appear close to the sun in the sky. Moreover, the heliocentric system dispensed with the variable motions introduced by Ptolemy, so Copernicus could even claim that he was trying to return to the uniformity that was a hallmark of Aristotle's physics.

Still, all of these arguments couldn't disguise the Copernican system's radical perspective on the place of humanity in the universe. By locating the sun at the center, Copernicus removed us from a specially privileged location and declared that logic trumped the needs of religion or philosophy. In what could be considered compensation, he placed us in motion among the celestial spheres, twining our destiny more closely with that of the stars. No longer was the earthly realm distinct from the ether. Now, studying the physical nature of the Earth offered the possibility of insight into the cosmic spheres, as both participated in the same circular dance. And while putting the sun at the middle shrank the sizes of the planetary orbits—the distance to Saturn in this new scheme was about forty million miles—the stars now had to be incredibly far away in order to appear motionless. In fact, the rotation of the Earth eliminated entirely the need for a finite sphere of fixed stars rotating once every twenty-four hours. It was quite possible, Copernicus reflected, that the universe could be boundless.

Centuries earlier, Saint Augustine had warned that the church should not endorse theories of the material world, lest it find itself on the losing end of the argument. All the same, prominent Catholic theologians and Protestant leaders, including both Pope Paul V and Martin Luther, argued against Copernicus. The church was hardly alone in rejecting any change in scientific tradition. Aristotelian philosophers recoiled from the Copernican system, and Tycho Brahe, probably the greatest observational astronomer in the pretelescopic era, dismissed it as absurd. They found it hard to let go of the spiritual and philosophical certainty of a known scientific system. Copernicus remained true to the cult of the circle, but his view of the universe contained a baffling new mystical element that we would now call inertia. In the heliocentric model, the Earth is constantly on the move. What propels it? And why then don't we all fly off? By setting the Earth in motion, Copernicus eliminated the heavenly ether as the propulsive force. He had to assume that some unknown factor keeps the universe running in smooth harmony.

Despite the reactionary responses to Copernicus, the scientific assault on the cosmos advanced sharply within a single lifetime. In 1609, Galileo Galilei turned his crude spyglasses skyward and witnessed sight after sight that didn't accord with the church-sanctioned cosmology. His discoveries are now part of every student's education, a triumphant retelling of the rise of the sci/religious faith during a time when it was surrounded by nonbelievers. Galileo saw that Jupiter is attended by four little stars—the satellites Io, Europa, Ganymede, and Callisto—that clearly follow paths around it, not around the Earth. Venus shows a cycle of phases like those of the moon, something that could happen only if Venus orbits around the sun, not the Earth. Furthermore, the sun has dark spots and the moon is covered with craters, physical flaws that undermined the alleged perfection of the heavens. Old-time religion faced a serious threat: a clever, outspoken, prominent thinker who was armed with a telescope.

Galileo's observations forced him to set aside his early doubts and proved to him that the Copernican system was correct. In fact, he became something of a Copernican zealot. His vehement attacks on the "Peripatetics" who did not believe in the motion of the Earth stirred up enmity within the church and ultimately led to the banning of *De Revolutionibus* in 1616. He stirred the pot further in his famous 1632 *Dialogue*, a mock discussion in which the supporter of the heliocentric system clearly gained the upper hand while the conservative doubter, tellingly named Simplicio, came off as something of a buffoon. The Vatican, unamused, ultimately called Galileo before the Inquisition, where seven of the ten cardinals sitting in judgment decided against him. Ironically, Pope Urban VIII was a former friend who had invited Galileo to write his book in part to demonstrate that the church was not repressing intellectual inquiry in Italy. Despite Galileo's follies, the church ended up looking increasingly irrelevant, and these episodes ultimately fostered more liberal religious interpretations that could accommodate the new astronomical ideas sweeping across Europe.

While Galileo undermined Aristotelian cosmology from the observational side, the German astronomer Johannes Kepler undermined its mechanisms. Kepler, fastidious in his thinking and his habits—today we might call him neurotic—performed a meticulous analysis of the motions of Mars that had been compiled by his mentor and tormentor, Tycho Brahe. So it happened that Tycho, who railed against Copernicus, indirectly spawned an even more extreme cosmological revision. Kepler fully embraced the Copernican system but found the discrepancies between the predicted and actual positions of Mars intolerable. He also hated the thicket of spheres on spheres and epicycles on epicycles. After a manic search for harmony, Kepler did a shocking thing: he replaced circles with ellipses, elongated shapes somewhat like the outline of an egg. Immediately all the observations fell into place and the solar system adhered to what he considered a much purer geometrical simplicity.

Continuing this theme, Kepler associated each planet with a perfect geometric solid. When he pictured how these solids would fit if one were nested inside another, he believed he could account for the observed spacing of the planets. Geometry functioned as the mystical element that kept order in his universe; he treated the concentric solids both as physical objects and as metaphorical expressions of Christian theology, with the sun at rest in the middle, representing God the Father, the creator of motion. Despite its overtly religious inspiration, Kepler's math helped destroy the old spherical cosmology and thereby separate astronomy from church doctrine. Kepler quietly made another revision to his planetary system that further aided in this disentanglement. Early on he had proposed that the planets were moved by souls, continuing a tradition extending all the way back to Aristotle, but in later editions of his *Mysterium Cosmographicum* he rejected this animistic interpretation. He recognized that the speed at which planets move varies according to their distance from the sun, which hardly seemed the attribute of an independent soul. So he switched from a spiritual to a material explanation: "When I considered that this moving cause weakened with distance, and that the sun's light too is attenuated with distance from the sun, I came to the conclusion that this is some kind of force. . . ."

The demise of the spheres and the corruption of the celestial orbs opened up a whole new domain to scientific inquiry. As long as planets existed on their crystal shells composed of unearthly elements, mingling among angels and souls, the question of why the heavens moved was one for philosophers and theologians. Perhaps Aristotle was right when he asserted that it is the nature of ether to proceed in perfect circles, with the whole system set in motion by an Unmoved Mover. Even the Copernican system could possibly work this way. But Kepler's free-range ellipses were another matter entirely. They dispensed entirely with the planetary spheres, so the old appeal to divine circular motion made no sense. Kepler proposed instead that some kind of force keeps the planets moving in their paths. This idea necessarily introduced a new question. What kind of force can reach through space and cause elliptical orbits? Kepler had no convincing answer.

But Isaac Newton did. The iconic British scientist solved the problem of planetary motion and redefined the place of God in the universe by rejecting, at long last, the ancient Aristotelian divisions between heaven and Earth. Historian Richard S. Westfall once described Newton's intellectual might as something beyond normal human comprehension. Consider the agony many of us experienced as students when we attempted to master calculus. Newton *invented* the foundations of calculus one year out of college. His personality was a mix of arrogance, brooding privacy, and alchemical obsessions. The calculus he created remained hidden in a drawer for years because Newton had no desire to place himself in the public eye. Likewise, his greatest notion and his greatest book came about only because of a bitter dispute with a scientific rival and the ceaseless nagging of Edmond Halley, the British astronomer who studied the bright comet that bears his name. Without those prods, Newton might never have written the text that catapulted science toward its sci/religious destiny.

The book was the *Mathematical Principles of Natural Philosophy,* better known by its shortened Latin title, *Principia,* published in 1687. The notion was the law of universal gravitation. The familiar tale that Newton had a sudden flash of inspiration after being conked on the head with an apple was probably invented to help keep the attention of high school physics students. True or not, this tiny anecdote contains a concise summary of what is so revelatory and downright magical in Newton's mathematical description of gravity. It is not just a practical explanation of the motion of a cannonball through the air or of that apple falling from the tree. It is also a manifesto of cosmic interconnectedness. Universal gravitation extends forever, so the force that pulls on the apple also holds the moon in orbit about the Earth and links one bit of the cosmos with every other. Gravity, like God, touches every piece of creation. Understand gravity, and God is pushed to the extremes, as a Creator and as a moral force, but not as an active participant in physical reality. This is Newton's other apple: the fruit of the tree of knowledge that sends man out of God's protected acres.

Universal gravitation provided a desperately needed replacement for the ether and Aristotle's crystalline spheres. It also made explicit what was implicit in Copernicus's revolutionary cosmology: The laws of heaven are the laws of the Earth, and vice versa. When Newton applied his theory of gravity to the planets, he found that they naturally yielded elliptical motion and the laws of planetary motion that Kepler had found. Newton's equations also allowed a universe of any size. The force of gravity falls off in proportion to the square of the distance, but there seemed no ultimate limit to its range. The pull of the planets held on to their moons; the sun's powerful attraction, in turn, kept the planets in line. The process could keep going to larger and larger scales without end.

Newton showed the way to banish the angels from the observable universe, and so divorce cosmology from theology.

For all his successes, Newton never managed to pin down the ultimate nature of the spooky attraction that could travel unfettered through empty space and pull disparate worlds together. "I have not been able to discover the cause of those properties of gravity from phenomena, and I frame no hypotheses; for whatever is not deduced from the phenomena is to be called a hypothesis, and hypotheses, whether metaphysical or physical, whether of occult qualities or mechanical, have no place in experimental philosophy," he wrote in *Principia.* And we now know that Newton's understanding of gravity was not complete. It is hopeless for explaining extreme phenomena such as black holes. Yet he attained such a close approximation of reality that his formulas are still good enough to send men to the moon or to measure the mass of a distant galaxy. More than two centuries passed before Einstein, the high prophet of modern cosmology, managed to improve on Newton's work and bring science to the pinnacle of sublime knowledge. Einstein's general relativity linked gravity, space, and time into a seamless whole. This connection completely erased Aristotle's division between earth and ether, and the religious division between heaven and hell. In Einstein's cosmology, we are unified with the spirit of the universe.

Many of Einstein's beliefs about the religious spirit of science grew directly out of his predecessor's ideas and attitudes. Newton placed great value on economy of explanation: "Truth is ever to be found in the simplicity, and not in the multiplicity and confusion of things," he wrote. And he saw his science as a kind of divine inquiry. He viewed the cosmos as the handiwork of God, and God as the ultimate gravitational "cause" that had eluded him. Formulating the rules of gravity, then, allowed Newton a small window into the mind of the Creator. He attempted to exploit his scientific ideas in order to know the size and form of the whole universe, trusting that the physical truths that hold here must hold everywhere. Newton, like Einstein, found himself responsible for the well-being of the universe as a result.

Newton believed that gravity acted instantly across any distance. In the post-Copernican world, it was clear that those distances could well be endless. Elliptical orbits were the order of the solar system. What, Newton wondered as if he were chatting with God, is the order of the universe as a whole? Initially he pictured the cosmos as a finite cluster of matter, encompassing ourselves and all the stars that we can see, surrounded by an infinite void. The unbounded extent of the universe was an essential part of Newton's world-view, because he deeply believed that the Lord must be infinite and eternal, "existing always and everywhere." After completing the *Principia* in 1687, however, he began to suspect that his clustered universe would not be stable. Gravity would pull and pull until every part of the cluster accumulated into a single mass; it might take a great deal of time, but eventually it would happen. This premonition of collapse was the first hint that the entire universe might change over time.

To Newton, the concept of an expanding or contracting universe was unthinkable. Cosmic order was a manifest expression of God's flawless creation. Yet the disruptive reality of gravity was evident everywhere. In the solar system, for instance, the planets all pull at one another, distorting the elliptical orbits with messy secondary motions that he feared could make the whole setup unstable. (In recent years, a number of scientists have shown that many solar system motions are indeed chaotic, fundamentally unpredictable over long periods of time. Fortunately Newton wasn't around to hear.) For the planets, Newton speculated that God might step in from time to time and clean up any discordant motions. But it seemed absurd that God would have created an entire unstable universe.

As far as Newton had moved past Aristotle, he still could not conceive of the universe as a dynamic whole. But Aristotle had the luxury of simply invoking the Divine in his claims about the ether. Newton had to figure out how to squeeze God into his testable equations. The basic problem was that gravity attracts everything to the center of mass, he decided. We are drawn to the Earth's center and hence weigh a little less on a mountaintop than at the seashore. The Earth in turn is drawn to the sun's center, and so on. As long as the universe has a center, it seemed, everything would be drawn toward that point and there could be no rest.

Newton was hardly the only scientist to reject the possibility of an evolving cosmos. Philosophers had imagined the universe as finite or infinite. In the eighteenth century, the British theologian and amateur astronomer Thomas Wright even imagined the possibility that our galaxy is just one of many island universes floating in the immensity of space. But prior to the second decade of the twentieth century, nobody had ever considered the possibility of an expanding or contracting universe. Such change seemed heretical, for either direction implies a beginning or an end. Although the Book of Genesis tells of a moment of creation, the idea of a literal beginning of time or an ultimate end was philosophically unappealing even to theologians such as Saint Augustine, who imagined God as timeless. An eternal cosmos is inherently more attractive. People may be weak and mortal, but there is some solace in believing the universe is indestructible and eternal.

In a series of letters with his friend, a young clergyman named Richard Bentley, Newton devised a cunning escape from this cosmic predicament. He abandoned his original picture of the cluster universe and proposed instead an infinite universe in which mass is uniformly distributed throughout. At any point, a mass would be drawn in all directions equally. The universe could not collapse toward its center, because it has no center. Instead it would gather in clumps to form stars, "scattered at great distances from one another through all that infinite space." The unstoppable nature of gravity created the problem. So Newton appealed to one intangible, the infinite extent of space, to undo the mischief created by another, the infinite reach of the strange force called gravity.

This argument is the direct predecessor of Einstein's Lambda, which also created an equilibrium that was supposed to counter the destructive pull of gravity. It is a clever way to build a static, eternal universe. Newton considered infinity a natural attribute of God and probably thought, incorrectly, that appealing to infinity would get him out of his bind. Alan Guth at the Massachusetts Institute of Technology, whose theoretical innovations helped reinvigorate the big bang model in the 1980s, sympathizes with Newton's error. Guth added a major new spiritual element into cosmology—an early episode of rapid expansion called "inflation"—to introduce a necessary balance into the big bang. "The failure of Newton's reasoning is an illustration of how careful one has to be in thinking about infinity," he writes, but then adds sternly, "From the modern viewpoint, an infinite distribution of matter under the influence of Newtonian gravity would unquestionably collapse." This turnabout was caused not by new discoveries but by new ways of looking at infinity and, by extension, new ways of looking for God.

In an infinite universe, the whole thing can collapse even without having a central point. Every part would fall toward every other part, from every direction, at an ever-accelerating pace. Ironically, what Newton conjured up was a system that would be the exact opposite of the real universe as we now know it, in which galaxies are fleeing in every direction without a center to the action. An unfortunate soul trapped in Newton's universe instead would see the galaxies rushing inward from all directions. The farther away a galaxy is, the faster it would approach, as the whole headed for a catastrophic crash-up. It would surely be a spectacular, terrifying sight. But it is one we will never witness. Newton's attempt to find a solution in infinity, like Einstein's later reliance on

Lambda, was the wrong solution, even though it ended up nudging sci/religion toward the right one (or at least a *more* right one).

Ever shrewd, Newton recognized the precarious nature of his solution. As long as the stars were distributed evenly throughout his infinite cosmos, he might imagine that the equal gravitational attractions in all directions would cancel out one another. The motion of even a single star would disrupt the balance, however. Einstein ran into a very similar problem trying to stabilize his model of the universe with Lambda. Newton ultimately evaded his predicament with a weak appeal to the Creator, who had thoughtfully placed the stars far apart, "lest the systems of the fixed Stars should, by their gravity, fall on each other mutually."

Newton had no other way out. Abandoning his phenomenally successful theory of gravity would have been absurd. Given his theology, however, abandoning the idea of an eternal, static universe would have been equally mad. What kind of Creator would bring the universe into existence only to fling it apart or dash it to bits? When infinity proved insufficient to rescue this model, Newton called upon the ultimate savior. The terms of the appeal reflected something new: Science was starting to dictate to God. Even though he unshakably regarded God as the ruler of the universe, Newton found himself in the position of specifying the placement of the stars so that his theory of gravitation would not be compromised. He thought he was being true to the old-time religion, yet Newton was sowing the seeds for the sci/religion that would replace it.

The great and terrible thing about looking to science for spiritual satisfaction is that its ideas are constantly open to criticism and contention. There is no guarded dogma available for protection. Newton had set down his best description of the cosmic laws of gravitation and started to treat the extent of the universe as a mathematical problem. Now it was possible for others to debate his solutions and question his views on cosmic divinity. With this disruptive freedom came a new kind of joy. Science promises that its practitioners will attain a precious, progressive kind of revelation: an ever closer approximation to the true nature of physical reality. This spiritual journey can proceed painfully slowly. Newton's ideas regarding the universal effects of gravity remained largely unchallenged until the early twentieth century. But when they arrived, Einstein's general theory of relativity and the discovery of the expanding universe elevated the scientific spiritualism to a whole new plane.

One of the most powerful objections to Newton's infinite, eternal cosmos centered not on delicate interpretations of gravitational stability, but on a question so simple that it sounds almost idiotic: Why is the sky dark at night? In an unbounded universe, the sky would be filled with an infinite number of stars and so should be awash with light. Yet every evening, the sun sets and the night is as black as ever. Edmund Halley, champion of the *Principia,* rushed to craft a plausible explanation. In a pair of papers presented in 1721, he proposed a variety of solutions, primarily arguing that the light from the most distant stars is so feeble that it would "vanish even in the nicest Telescopes, by reason of their extreme minuteness." Halley considered this paradox important enough that he discussed it at a meeting of the British Royal Society, where the elderly Newton was in the audience. The stakes were high. Here was simple way to judge the magnitude of God's handiwork and evaluate Newton's belief that the glory of the Lord should be reflected in the infinite expanse of the universe.

Despite Halley's arguments, the question of the dark sky resurfaced from time to time in the following decades. It again came to the fore in the 1820s, when Heinrich Wilhelm Olbers, a German physician turned astronomer, revisited and popularized this astronomical mystery; as a result, it is now usually known as Olbers's paradox. Olbers exposed a gaping flaw in Halley's logic. If every line of sight eventually meets the surface of a star, it doesn't matter whether some of those stars are extremely far away. Every speck of the sky will be filled with starlight, and the whole should appear as bright as a hundred thousand suns. Olbers still believed in Newton's universe—"Is it conceivable that the almighty Creator should have left this infinite space empty?"—so he looked for a loophole and found one. Starlight adds up only if it passes freely through space. "This absolute transparency of space is not only wholly unproven, but also quite improbable," he wrote. Dark intervening clouds or interstellar mist must block the light from the most distant stars, he concluded happily, rendering them invisible.

As physicists came to understand the nature of radiant energy, however, they realized the futility of Olbers's solution. Energy from the most distant stars would not vanish. It would be absorbed and become a part of the obscuring clouds. If the universe were endlessly old, the energy accumulating in those clouds would cause them to grow hotter and hotter until they too glowed white hot, again as brilliant as the surface of the sun. We should be blinded and fried by an onslaught of infinite starlight, yet all is calm. The paradox stands unless one of Newton's infinites is wrong: the universe cannot be both infinitely old and infinitely abundant with stars. Finally, in the twentieth century, astronomers discarded both infinities in response to Einstein's equations and the discovery of the expanding universe. In this way they solved one set of philosophical issues but created others, most notably how it all began and how it all will end.

The solution to Olbers's paradox emerged slowly from efforts to move beyond Newton's vague description of stars scattered evenly throughout an infinite expanse and to develop a more concrete picture of the scale and organization of the heavens. Even in Newton's day, it was clear that his model was a broad simplification of reality. The shimmering band of the Milky Way slices the heavens in two, blatant evidence of cosmic asymmetry. When Galileo trained his spyglass on the Milky Way, he saw right away that it appears packed full of faint stars, like the lights of a distant port descried from far offshore. This gathering clearly betrayed an organized pattern to the universe, but it took scientists more than three centuries to understand fully what that pattern is.

Newton's theory of gravity, which explained how a conglomeration of stars could hold together by their mutual attraction, might plausibly have led him to conclude that the Milky Way is just such a starry swarm. In reality, Newton was a big-picture spiritual thinker who didn't get involved in fussy matters of astronomical cartography, much like Einstein after him. The first person to knit observation and logic into a coherent picture of our home galaxy was a man who knew a lot about logical thinking, Immanuel Kant. These days he is better remembered for philosophical innovations like the categorical imperative, but the lines between science and philosophy were not as clearly drawn in the eighteenth century as they are today. Kant started thinking about the structure of the Milky Way after reading (and slightly misunderstanding) a review of a book by Thomas Wright, in which Wright discusses the possibility that we live in an enormous disk of stars. Kant grew enamored of this notion and developed it far beyond Wright's brief and rather speculative description.

In his *Universal Natural History and Theory of the Heavens* of 1755, Kant argued that such a structure naturally arises from Newton's laws: "The influence of the fixed stars, as of so many suns . . . are striving to approach each other on account of this mutual attraction . . . sooner or later each implodes in a single clump, unless this cataclysm is prevented, as it is with the spheres of our planetary system, by the action of centrifugal forces." Extrapolating from the mechanics of the solar system, where the planets all orbit the sun in more or less the same plane, Kant deduced that the centrifugal forces would similarly arrange the spinning mass of stars into an enormous disk—an idea that he generously, even excessively, credited to Wright. When we look perpendicular to the disk, Kant explained, the sky appears dark and we see only occasional, scattered stars. But when we look along the disk, viewing it edge-on from within, we see all of the distant stars more or less lined up. The glow of the Milky Way in the sky is "a densely illuminated belt of innumerable stars aligned like the greatest of great circles."

Then Kant stripped back another layer of cosmic mystery by combining his analysis with the astronomers' increasingly complete census of the sky. Telescopic observations had revealed dozens of fuzzy patches of light, known generically as nebulae (Latin for "clouds"), dotting the sky. These nebulae, he proposed, might be other systems like our own, scattered through the tremendous depths of space. "Their shape, which is just what it ought to be according to our theory; the feebleness of their light, which demands an assumption of infinite distance—all these correspond perfectly with the assertion that these elliptical figures are just such world systems and, so to speak, Milky Ways, whose structure we have just unfolded," he wrote. From Newton's laws Kant understood, in at least a general way, that these systems should exhibit collective circular motions like the orbiting of the planets. Many of the nebulae have a whirlpool-like appearance, which Kant took, quite correctly, as a sign of that motion. His conclusion that the Milky Way is but one among innumerable galaxies allowed him to return to an unbounded, Newtonian cosmology in which "the whole infinite extent of its greatness is everywhere systematic and interrelated."

Intrigued by these ideas, the masterful German British astronomer William Herschel set out to determine if Kant's fanciful musings could survive a good thrashing from the scientific doctrine of falsification through observation. Using giant telescopes of his own design, Herschel performed a detailed census of the sky and attained vastly improved views of the enigmatic nebulae. Herschel was a tireless worker who kept careful records as he picked over every degree of the heavens. His efforts paid off spectacularly with the 1781 discovery of Uranus. This triumph, the first addition to the solar system since antiquity, delighted King George III, who soon after made Herschel the royal astronomer. (This was during the American Revolutionary War, when George III could use some good news, and before the full progression of his porphyria, the hereditary blood disease that tainted the king with a lasting reputation as an irascible eccentric.) With this newfound support, Herschel set off to map the "Construction of the Heavens." He made the first stab at measuring the Milky Way by searching for the faintest stars that must lie at its edge. This overly ambitious task never yielded the answer he sought, but it did confirm Kant's notions about our galaxy's disklike form.

During his long hours at the eyepiece of the world's most powerful telescope, a forty-foot-long brute with a forty-eight-inch metal mirror at its heart, Herschel familiarized himself with the different kinds of nebulae and recognized that they vary tremendously. At first he believed all of them were composed of stars. But by 1791 he had changed his mind and realized correctly that the nebulae fall into two broad categories: those that are gatherings of stars so remote that they all blend together, and those that are truly gaseous or, using his terminology, composed of "luminous fluid." This fluid might "produce a star by its condensation," he speculated, introducing the idea that celestial bodies might evolve even if the universe as a whole is immortal. Herschel also recognized objects such as the Andromeda nebula for what they truly are, distant galaxies as imposing as our own. Their feeble glow hinted at the tremendous extent of space between them and us. Herschel affirmed the Newtonian belief in "the indefinite extent of the sidereal heavens, which must produce a balance that will effectually secure all the great parts of the whole from approaching to each other," but he added a tremendous appreciation for the complexity of how these parts are arranged.

Despite his best efforts, Herschel could not quantify the dimensions of the whole. He could only guess, because nobody had yet found a way to measure the most basic unit of the cosmic yardstick, the distance to the nearest stars. Herschel died twenty-six years before the problem was solved by Friedrich Wilhelm Bessel, a German astronomer who had studied under none other than Heinrich Wilhelm Olbers. Bessel took Aristotle's old argument in favor of an immobile Earth and stood it on its head. As the Earth circles the sun, its changing position must make nearby stars appear to shift back and forth against the more distant stellar backdrop. This perspective effect, called "parallax," is an essential part of our depth perception. Place a vertical finger in front of your face and close first one eye, then the other. The way your finger jumps relative to the background is parallax, and it works exactly the same for stars as for fingers. If you know how much the Earth's location changes over the course of the year and how much the apparent position of a star has changed over the same time, high school trigonometry will give you the star's true distance.

Because nobody had succeeded in observing such a parallax, Bessel knew the size of this motion had to be extremely small. He sifted through data on fifty thousand stars, searching for the most promising one to study. Finally he settled on 61 Cygni, a yellow orange double star barely visible to the eye in the constellation Cygnus, which he correctly deduced must lie relatively nearby. His first set of observations in 1815 went nowhere. Two decades later he tried again, armed with a vastly superior measuring telescope known as a heliometer. Over a year and a half Bessel watched 61 Cygni, and this time he clearly saw the star shift back and forth by one-third of an arc second—about the apparent angular size of a quarter seen at a distance of ten miles. In 1838, Bessel announced that the distance to 61 Cygni is "657,700 mean distances of the earth from the sun," or 10.3 light-years, very close to the modern value of 11.4 light-years. One light-year is about six trillion miles, so outer space is huge and unfathomably empty. Whereas great thinkers had formerly resorted to counting up crystalline spheres, astronomers now had a ruler with which to measure the heavens.

But by the end of the nineteenth century, the human exploration of the heavens had reached an impasse. Our measuring sticks weren't long enough to assess the extent of God's domain. In fact, parallax gave accurate results only for the closest stars. Beyond that, the effect became imperceptibly minute. It gave only vague hints about the extent of our galaxy, and it told nothing about the spiral nebulae that Herschel believed were mighty whorls of stars in their own right. Without that information, speculation about the nature of the spiral nebulae remained speculation alone. The situation changed little for the better part of a century: at the end of World War I, astronomers were still engaged in a debate about the scale and structure of the universe. Newton had formulated mathematical laws whose reach seemed infinite, but at the end of the nineteenth century his successors were still struggling to understand the lay of the land in their galactic backyard. What lay beyond was a mystery, a no-man's-land denied to science. Astronomers were similarly limited in their ability to explore across heavenly time. By declaring the cosmos eternal, Newton essentially declared the whole question of origin and first causes as off-limits to scientific inquiry. His theory of universal gravitation didn't lead to universal enlightenment.

Undeterred, scientists carried out their quest for the spiritual by investigating mysteries closer to home. They couldn't trace the origin of an eternal universe, for instance, so they started trying to measure the age of the Earth. In the Book of Genesis, the origin of the Earth and the origin of the universe are part of the same, week-long episode of divine creation. During the medieval and early modern periods, Catholic theologians had investigated the biblical chronology in fine detail. Saint Augustine examined the rate of progress in human history and concluded that the beginning of the world—but not, of course, the beginning of God—occurred around 5000 B.C.E. The most notorious age measurement came from James Ussher, the archbishop of Armagh, Ireland, in 1650. In an unwitting act of reductio ad absurdum, he placed the creation of the Earth on the evening before October 23, 4004 B.C.E. The growing power of science meant that the Earth's history could now be submitted to a different kind of calculation.

Herschel's sky surveys suggested that stars might condense from clouds of luminous gas. Later, in the 1840s, the German philosopher and scientist Hermann von Helmholtz formulated the concept of conservation of energy and argued that the sun's heat must derive from gravitational energy that is liberated by its contraction. In other words, the sun shines because it is constantly falling in on itself. This effect—which in fact does apply to young stars in the first stages of formation—offered a way to calculate the age of the sun and, by extension, of the Earth. If the sun began as one of Herschel's clouds, then it was a simple matter of mathematics to determine how much gravitational energy resides in such a cloud and how long that energy could keep the sun shining. Helmholtz's friend Lord Kelvin, one of the most influential scientists of the latter half of the nineteenth century, followed through on this work. In 1863 he declared the Earth to be less than two hundred million years old, a number that he steadily revised downward in later years.

From a cosmological standpoint, the crucial point is not that Kelvin was wrong, but that he so boldly insisted the age of the Earth could be determined through scientific inquiry. And although he never presumed to speak about the age of the universe, his increasingly emphatic calculations of when our planet formed undermined the Newtonian belief that God's glory is reflected in the permanence of his creation. Helmholtz, meanwhile, began looking in the other direction and staking science's claim on the future. Working from Kelvin's ideas about the dissipation of heat,

Helmholtz anticipated that all of the energy powering the stars would someday be depleted. He spoke despairingly of the endless decline "which threatened the universe, though certainly after an infinite period of time, with eternal death." But he still did not dispute its eternal existence.

In the 1850s, just as Helmholtz was bemoaning our cosmic future, two other researchers were testing the limits of another of Newton's signature concepts, the cosmic uniformity of natural law. Newton's equations of gravitation seemed to apply anywhere in the universe, contradicting Aristotle's old notion that the ether follows different rules from those governing the common elements. Now the German physicists Gustav Robert Kirchhoff and Robert Bunsen had found a way to see if celestial objects are composed of the same elements, built from the same kinds of atoms, as the Earth. Earlier research had shown that when sunlight passes through a fine prism, the resulting spectrum is full of thin dark lines; conversely, elements heated by a carbon-arc lamp produced thin spectral lines of light in an analogous pattern. Kirchhoff put the pieces together and concluded that the dark absorption lines in sunlight are just the emission lines seen in silhouette. Together with Bunsen, he started examining the solar spectrum for the spectral fingerprints of familiar elements. Within a few years they identified thirty common elements, behaving in the sun just as they do on the Earth.

Starting in 1862, Sir William Huggins, a British astronomer blessed with enough wealth to afford his own observatory, extended this work to the stars and found the same atomic absorptions there as well. When he trained his telescope on the nebulae, he again saw the chemical signatures of known elements, but this time as emissions, not absorptions—a distinction showing that the nebulae consisted of thin gas at low pressure. "The answer, which had come to us in the light itself, read: Not an aggregation of stars, but a luminous gas," he recalled excitedly. This finding later caused some confusion, as it obscured the distinction Herschel had made seventy years earlier and seemed to suggest that all nebulae are nothing but wisps of vapor. For the moment, however, these results were both revelatory and reassuring. "A common chemistry, it was shown, exists throughout the universe," Huggins wrote. Newton's faith in uniformity was vindicated. The reach of the astronomers extended again, such that they could reply to the poets and tell them what stars are made of.

Yet Newton didn't lead science into the promised land, nor did Huggins. Both men suffered from limits of vision. For Huggins the problem was literal: he could not study what he could not see, and spectroscopy demanded a lot of light. For Newton the problem was too much of that old-time religion. He could not bring science into the farthest reaches of space because he was shackled to his faith in an infinite universe and absolute space, which he regarded as necessary attributes of an omnipotent God. Albert Einstein, Newton's heir and one of his few intellectual rivals, did not carry this theological baggage. Newton had to rely on God to provide a reference against which to measure time and location; Einstein derived equations that gave shape to space all by themselves. Newton's theory of gravity required divine intervention to keep it both endless and eternal. In seeking a more comprehensive description of gravity, Einstein redrew the universe, melding Newton's devotion to inviolable, all pervasive laws with the geometric beauty of Aristotle's spheres. This new universe was finite, spherical, static, and eternal. It was the heavenly vision that heralded the birth of the new faith, sci/religion.

In creating his radical cosmology, Einstein stitched together a rational mysticism, drawing on—but distinct from—the views that came before. Galileo had tried to define a line of demarcation between science and religion by insisting that the Bible tells how to go to heaven, not how the heavens go. Einstein endlessly violated that boundary by redefining God both as an ally and an end in his search for scientific truth. "God does not play dice with the universe," and "the Lord God is subtle, but malicious he is not" are just the most famous of Einstein's sci/religious declarations. His deity was not the interventionist God of Abraham and Isaac, but something more complex and abstract—not so much the Creator of the universe as the embodiment of a beautiful and economical set of physical laws. Einstein snatched the deist God of Spinoza and impressed Him into duty for science. This God does not dictate to His human subjects. Rather, they dictate to Him through their scientific investigations. Einstein built on the process begun when Newton insisted that his Creator would take care to place the stars sufficiently far apart to avoid gravitational disruption. God embodies cosmic law, and cosmic law is revealed through what comes out of the equation and the eyepiece.

Einstein believed the path to God lay in understanding the way the universe works today, not how it began. Initially he didn't consider a beginning at all, again following in the tradition of Newton and Aristotle, yet his ideas provided the underpinning for the grandest, most persuasive creation story ever told. First came the big bang theory. More recently, cosmologists have started proposing an endless succession of bangs and, in an echo of Saint Augustine, arguing that the beginning of our universe was not necessarily the beginning of time. What all the modern scientific thinkers have in common is a faith that scientific inquiry lives where religion once ruled, back to the moment of creation. They are all driven by a holy conviction that they are on a quest toward absolute cosmological truth and that mathematical unity, consistency, and beauty will lead them there. They all seek the missing elements needed to elevate the known physical laws into a comprehensive model of the state of the universe. They have even resurrected Einstein's Lambda to perform that task. Ask them if they believe in religion and nine times out of ten they will profess a careful agnosticism. Examine their research, however, and it will leave no doubt that they worship in the Church of Einstein.

Of course, any attempt to understand this world necessarily has a subjective human element—at the very least, an element of faith in the comprehensibility of the natural world. How else could Eudoxus believe that his geometry would allow him to mimic the movements of the planets? How else could Newton believe that the same gravity that holds us onto our planet reaches to the farthest star? What keeps changing is the scale and the stakes. Copernicus, with some help from Galileo and Kepler, exiled us from the motionless center and opened the possibility that human inquiry could extrapolate from the Earth to the heavens. Newton codified that idea with his laws of universal gravitation. Einstein brought all of the universe—not just its function, but also its form—down to Earth in his field equations. The result is an inevitable theological turnabout, in which God began to take man's image. Even that devoted-sounding sentence "God is subtle, but malicious he is not" presumed that it lay within Einstein's power to infer or decree divine intent.

We've been headed this way since the days of the Greek academies. It was like an ancient prophecy that finally became clear when Einstein arrived: When we reach for vastness it is like grasping for the Divine. We are a small huddle of life clinging to a moist blue rock that swings endlessly around a modest yellow star on the outskirts of a galaxy floating among billions of others. We evolved to hunt for food, to make simple tools, to find mates and have sex. Along the way we also evolved consciousness and a desire to know, and it got wired deep in the brain along with our other instincts. We crave an understanding of our place in the world. Once we sought that through Scripture. Many people, in many parts of the world, still do. Increasingly, however, they are finding the ecstasy lies elsewhere. Einstein put God in physics and made physics into God.

Chapter Three

THE CHURCH OF EINSTEIN IS FOUNDED

IN THE EARLY WINTER months of 1917, Albert Einstein huddled in his bachelor's flat at Wittelsbacherstrasse 13, in central Berlin, and prepared to reshape our deepest insights and our everyday ideas about the physical workings of the world. Gallstones shot pain through his abdomen, and the war-reduced Prussian diet disagreed with his sensitive stomach, but his mind was elsewhere. Each stroke of his pen conjured up a strange new place where light bends like taffy and galaxies float atop a sea of curved space, like ships scattered across the global expanse of the Earth's oceans.

Einstein was no longer the boy genius who formulated his theory of special relativity at age twenty-six while laboring as a technical officer at the Swiss Patent Office in Berne. Now thirty-seven, he had achieved indelible fame within the scientific community, although he was still not quite a household name. Recognition did not make his labors any easier. For years he had struggled to broaden his theory of relativity into a set of equations that would explain the fundamental nature of gravity and the structure of space, a task that had eluded even the great Newton's intellectual grasp. He endured painful setbacks and repeatedly scrapped his failed formulations, but, like Job, he persevered. In 1916, Einstein finally delivered his general theory of relativity. He had taken another leap forward in his determined journey toward unlocking the secrets of the universe and thereby entering the mind of God.

This was, he wrote to physicist Arnold Sommerfeld, "the most exacting period of my life; and it would be true to say it has also been the most fruitful." Crafting general relativity was the biggest challenge he had ever tackled. He compared it, in a letter to his friend Paul Ehrenfest, to "a rain of pitch and sulfur." The work exhausted him, both physically and intellectually. But it led to one of his most profound discoveries. According to his new elaboration of relativity, gravity warps the structure of space. Newton had hit a dead end when he tried to understand how gravity links bodies through the emptiness of space, essentially relying on God to transmit the attraction responsible for action at a distance. Einstein hit on a new formulation: Space and matter are equal partners, with gravity the result of the interaction of these tangible and intangible elements. This result soon led to a second, inspired break with Newton, in which Einstein abandoned the infinite universe so that he could embrace all of space with his equations. Considering the universe as a whole, Einstein realized the collective warping would cause space to curve in on itself, forming a giant ball. Like the surface of the earth, this spherical universe was limited in extent but had no edge.

When he scrutinized cosmic implications of the theory, however, Einstein found one aspect of the result disturbingly wrong. The finite universe described by general relativity inclined either to expand, carrying all the stars away from one another, or to collapse disastrously under the force of gravity. Here Einstein remained rooted in the classical conception of Aristotle and its Newtonian echo: the universe is eternal and unchanging, how else could it be? So the great man shrugged, again took up his pen, and added an extra component—the cosmological constant, represented by the

Greek letter *lambda*—into the mathematics describing the way space and matter interact. Lambda gave empty space an outward pressure that could precisely negate the inward pull of gravity. Once again the universe was static, exactly the way God would have made it.

In this simple act, Einstein breezed past the tradition of Copernicus and Galileo, establishing a vastly expanded domain for science. Now its scope included not only all the observable phenomena of nature, but also the master framework of the universe within which all those phenomena occurred. And by embracing all of physical reality, Einstein was no longer content to have science praise the ultimate glory of God, as Newton had been. General relativity seemed ready to tackle everything that is out there. If so, science was no longer restricted to describing God's handiwork; now it could describe God. Einstein popularized this conception through his widely reported comments about the nature of God and about what God would or would not do. But the new attitude would have come across even without Einstein's proselytizing. Simply by putting Lambda into his relativity equations, he implicitly claimed that he could draw on the authority of modern science to discern the structure of the universe and, by extension, the nature of its Creator.

It was a long journey from Einstein's famously unpromising childhood to his adult role as shaper of the universe. He was born on March 14, 1879, to Pauline Einstein (née Koch) and Hermann Einstein in an unremarkable apartment in the German town of Ulm. His father was an optimistic but repeatedly unsuccessful entrepreneur who dabbled in ventures selling goose down, manufacturing boilers, and installing electric lamps—ventures that, for a while, at least, kept the family in a comfortable, middle-class existence in Munich, where the family moved when Albert was one year old. "Nothing in Einstein's early history suggests dormant genius," writes Ronald Clark, one of Einstein's most perceptive biographers. If anything, the signs pointed the other way. Stories of the young Albert as a backward child, poor at language, have become a standard part of the Einstein legend. They come predominantly from Einstein himself, who recalled his sluggish early development as something of a blessing in disguise: "The normal adult never stops to think about the problems of space and time. These are things which he has thought of as a child. But my intellectual development was retarded, as a result of which I began to wonder about space and time only when I had already grown up."

The formation of Einstein's peculiar brand of deism seems easier to trace, although the amount of information about his early religious views is limited. His parents were Jewish, but they rejected most aspects of Jewish practice, and the young Albert attended Catholic school and lived in a predominantly Catholic region. At age eleven, he had absorbed sufficient mixed religious instruction that he stopped eating pork and composed private hymns. While he was studying at the strict Luipold Gymnasium, however, Einstein entered a rebellious phase and his thinking started to shift. Soon he was reading scientific texts and Kant's *Critique of Pure Reason.* Despite, or because of, his continuous religious instruction, he became convinced that "much of the stories of the Bible could not be true." He never had a bar mitzvah ceremony. Later in life he looked back at his childhood religious phase as "a first attempt to liberate myself from the 'only personal.' "

Yet Einstein did not abandon religion. Rather, he made peace with God, albeit a very different kind of God from the one he heard about at the Luipold Gymnasium. Years later, when he met with the Jewish philosopher Martin Buber, Einstein told him, "I want to know how God created this world. . . . I want to know His thoughts, the rest are details." Sometimes, as with the whole story of Lambda, it seemed that Einstein chose to dictate God's thoughts and then step back and see if the Lord agreed with his decrees. When he famously insisted that "God does not play dice with the world"—a line that was repeated, discussed, and rephrased many times during his life—Einstein was not interpreting a religious text, he was writing one of his own. He once exposed this method explicitly: "When I am judging a theory, I ask myself whether, if I were God, I would have arranged the world in such a way."

Einstein found his divine path through science and transformed science in the process. His journey down that path evidently began during his poorly documented teenage years. By 1895 he had completed his secondary education (leaving without a degree), applied to the Swiss Federal Polytechnic School, and sent his uncle Caesar a remarkable letter. In it, he included a handwritten draft of a paper entitled "Concerning the Investigation of the State of Ether in the Magnetic Field." The title refers to the then widespread theory of how waves of light move through empty space. The teen Einstein already exhibited a keen interest in understanding the propagation of light, a fascination that would later show up both in his theories of relativity and in his formulation of quantum physics. A decade later, Einstein's new physics put the ether theory to rest. Even as a brash eighteen-year-old, he felt the stirrings that would make him a great scientist and a great religious leader. "Strenuous labor and the contemplation of God's nature are the angels which . . . will guide me through the tumult of life," he declared.

While young Einstein was pondering how to carry out this project, a number of scientists were smugly speculating that there might not be too much left to contemplate, now that their studies had succeeded so well in describing the mechanics of the world. The eminent American physicist A. A. Michelson summed up the spirit of the times: "The most important fundamental laws and facts of physical science have all been discovered, and these are now so firmly established that the possibility of their ever being supplemented in consequence of new discoveries is exceedingly remote." This was by no means a universal view, but many otherwise sensible people believed there was little need for new theories, because the old ones were doing the job quite well. Science always moves forward when challenged. At the end of the nineteenth century, however, the prevailing feeling was one of intellectual and spiritual stagnation. The gaps in understanding hardly seemed sufficient to spark a fundamental realignment between science and religion. As often happens, the magnitude of the problems became clear only in hindshight.

Newton's old cosmological predicaments remained. Olbers's paradox still had no satisfactory answer; the universe somehow managed not to come crashing down all around. There was also a more concrete, albeit much smaller, inconsistency in his theory of gravity. Newton's equations implied a clockwork universe that operated with unerring mathematical precision. Yet Mercury, the planet closest to the sun, did not behave as expected. Its highly elliptical orbit does not stay put. The direction of the long axis of the orbit rotates gradually around the sun, so that the point where the planet is farthest from the sun sweeps out a huge circle. The pattern of successive orbits marks a kind of flower petal pattern, similar to the lines produced by a spirograph set. Newtonian gravitation could account for most, but not all, of this motion. The error was so small that it would take 180 million years for it to produce one additional circuit of Mercury's orbit. But it was undeniably there, and none of the proposed fixes totally eliminated the problem. God's clockwork universe was not keeping perfect time.

Other pieces of Newton's physics, when scrutinized closely, seemed incomplete as well. He had never explained the fundamental nature of gravity and especially had never confronted the problem of how gravity propels itself through empty space. By the end of the nineteenth century, scientists thought they had a solution to this mystery of action at a distance, at least for the case of light and other forms of electromagnetic radiation. Their standard response was that an empty vacuum is never truly empty. All of space is permeated with an unseen and unfelt material, known as "the lumiferous ether," that acted as a kind of cosmic signal relay. When light streaks across the seeming nothingness between the sun and the Earth, it is actually vibrating through the ether like waves traversing the ocean. Perhaps gravity could propagate through space in the same manner. Never mind that nobody could explain what ether consisted of or provide clear evidence that it existed. The modern ether, like the one invoked centuries earlier by Aristotle's, served a philosophically crucial purpose, giving form to the cosmic void.

Yet, also like Aristotle's ether, the updated version bore no resemblance to any known or imaginable earthly substance. It extended terrestrial logic into space but at the same time built a clear distinction between our world and the world above. The luminiferous ether needed to be stiffer than steel but completely invisible and imperceptible. Two physicists working together in the United States—first Michelson, the fellow who eventually declared science was petering out, and, later, Edward Morley—put the ether to a test. As the Earth moves through space, it must plow through the ether. Light should therefore appear to move at different speeds in different directions, faster in the direction where the ether is blowing into our faces and slower where the ether wind is at our backs. Michelson and Morley set up an experiment at Western Reserve University in Cleveland that split a beam of light into two, sent the halves in perpendicular directions, and then recombined them. One beam should have moved faster through the ether than the other, in which case the two beams would be visibly out of alignment when they came back together. In 1887, Michelson and Morley announced their results: They found nothing. Michelson repeated the experiment at different latitudes in case the ether was being dragged along by the body of the Earth. Still nothing.

This negative result was the first strong indication that the observed speed of light can never change, which Einstein later fashioned into a guiding principle of physics. It also cast a serious blow against Newton's belief in absolute space, an infinite, invisible universal reference grid that he referred to as "the Sensorium of God." Ether theory attempted to reconceive this immovable backdrop as a material accessible to science, but the attempt to reach it by experimental means failed miserably. Einstein probably didn't learn the details of the Michelson-Morley experiment until years later, possibly not until after he had developed his basic ideas about relativity; his later comments are ambiguous and contradictory. But his studies at the Swiss Federal Polytechnic School certainly introduced him to the work of Scottish physicist James Clerk Maxwell and Austrian scientist and philosopher Ernst Mach, two brilliant minds whose work helped bring about Einstein's relativity theory and the death of the ether.

During the 1860s, Maxwell formulated a set of equations describing light as a wave in the form of an oscillating electromagnetic field. He also analyzed the flow of heat as a function of statistical changes in the motions of the molecules in a gas. Einstein's expansion of those ideas blossomed into a thoroughly new understanding of the nature of light and the rules that govern it. Mach's thinking had an even more profound effect on Einstein. Mach is best remembered for his research into the physics of supersonic motion, which is why faster-than-sound speeds are described on a Mach scale. But he was also a cunning philosopher of science who argued that our notions of space and time are fundamentally limited by the human senses we use to understand them, a line of reasoning that reappeared, wonderfully mutated, in the general theory of relativity. Mach was such an empirical purist that for years he refused to believe in the reality of atoms. In *The Science of Mechanics* he railed against Newton's "conceptual monstrosity of absolute space," an attack that deeply influenced Einstein when he read it as a college student. Particularly in his early years, Einstein described himself as a devout Machist and cited Mach as the inspiration for his 1917 cosmology—even though, by introducing Lambda into the equations, Einstein defied his idol's warnings against any scientific idea that cannot be derived from experience.

By the closing years of the nineteenth century, physics was astir with so many strange new discoveries that even the most sheltered thinkers could no longer cling to the notion that science had achieved a full explanation of the natural world. Wilhelm Roentgen encountered an unknown, penetrating kind of radiation he called "X rays." Henri Becquerel discovered radioactive decay in uranium salts. J. J. Thomson uncovered evidence that the atom—assumed indivisible since it was first conceived in the philosophy of Leucippus and Democritus in the fifth century B.C.E.—contains a smaller particle, the electron.

Young Einstein started to develop the outlines of his cosmic religion amid this intellectual ferment. He graduated from the Polytechnic in 1900 and set out in search of an academic position. After months of fruitless searching, he landed a job at the Swiss Patent Office in June of 1902. The disciplined setting forced him to express himself clearly. More important, the rote nature of the work and the quiet evenings and Sundays it allowed gave him time to work out his ideas. Had he been in an academic setting, Einstein reflected some fifty years later, he would have felt pressure to churn out scientific work even if it was mediocre. In the shelter of the patent office, he could let his mind wander. He enjoyed the encouragement of a co-worker and former classmate, Michele Angelo Besso, who became a lifelong friend and confidant. This was also a time of tremendous personal change for Einstein: his father died in October 1902, and he married Mileva Maric in January of the following year.

Like Moses, Einstein had spent his time in the desert, and now he was ready to answer the divine call. Historians call 1905 Einstein's "annus mirabilis," his miracle year. This was his moment of greatest intellectual ferment, although the key insight that led him to the scientific promised land was still years away. In rapid succession he published four papers in *Annalen der Physik*—three of them in a single volume—that transformed the face of physics. He managed this feat even though he was then on the far fringes of academia. Einstein's access to current scientific debates came primarily through the physics journals. He was not in contact with the major researchers of his day, nor did he have access to the results from the newest experiments. What he did have was a keen sense that physics could be rebuilt in a way that was more harmonious, that looked better, and that he therefore felt must be true. In that sense he had access to the same tools that had aided Copernicus and Newton, a logical devotion to the principles of economy and unity in understanding the world.

One of the 1905 papers described the motions of molecules in statistical terms. A second worked out an explanation for the photoelectric effect, a phenomenon in which electricity flows from certain substances when they are exposed to light. Einstein found common ground in these two papers. He realized that the way light strikes a surface is analogous to how gas molecules hit the wall of a box, an idea that forms the foundation of quantum physics. But the main attraction of the 1905 papers was Einstein's formulation of his theory of relativity. More precisely, it was his special theory of relativity, so named to distinguish it from the general version he developed a dozen years later. "Special" in this context means that the theory applied only to restricted situations in which gravity and acceleration were not important factors.

The paper's title, "On the Electrodynamics of Moving Bodies," hardly suggests its revolutionary content. In the span of nine thousand words, Einstein overturned three hundred years of wisdom regarding how to determine an object's motion or how to define the timing of an event. Much as Copernicus put the sun at the center of the solar system to simplify the motions of the heavens, Einstein set out to remove "asymmetries that do not appear to be inherent in the phenomena." For instance, moving a magnet across a piece of wire generates current, but the equations describing the motion of a magnet past a wire were different from those describing the motion of a wire past a magnet. Einstein proposed a radical way to remove this arbitrary distinction: Abolish Newton's Sensorium and with it the entire notion of absolute motion and absolute space. The old universe and whatever we thought we knew about who oversaw it were dead.

Hundreds of books and articles, including Einstein's own, have attempted to translate the concepts of relativity for the lay public, yet the theory remains stubbornly counterintuitive. A man is standing in a train car moving at ten miles per hour. Suddenly he starts to run ten miles per hour in the same direction as the train's motion. How fast is he moving? It seems like a trick question. Anyone trained in classical mechanics or everyday common sense knows the answer must be twenty miles per hour. Einstein too saw this as a trick question, but he perceived that the trick is to ask, "Moving relative to what?" Relative to the train, the man is running only ten miles per hour. But even relative to the ground, Einstein argued, he is actually traveling a bit less than twenty miles per hour. At such low speeds the effect is minute, but it becomes huge when the speeds approach the velocity of light. Assume now that the train is moving at nine-tenths light speed, and the runner sprints forward at another nine-tenths of light speed. From the ground, the combined motion still appears distinctly less than the speed of light.

The reason for this counterintuitive kind of addition is that the speed of light is a fundamental limit in the universe. Einstein realized that theory and observation alike declare the physics of light appears the same regardless of how the observer is moving. For this to be true, velocities cannot keep adding up they way we naively expect. A beam of light moves at the speed of light no matter what, even if it comes from a flashlight aboard that superfast train. To the passenger aboard the train, the beam must recede at the familiar old speed of light. To a train spotter watching the cars whiz by, however, the passenger must nearly be keeping pace with the beam of light—because from the train spotter's vantage, that light still travels in its usual way, despite the added motion of the train. The only way both observers can be correct is if time and space are variable things that depend on whose perspective you choose.

Relativity expresses Einstein's faith that the universe is governed by inviolable laws and that these laws are accessible to human investigation. In 1906, Einstein wrote that "a theory is the more impressive the greater the simplicity of its premises is, the more different kinds of things it relates, and the more extended is its area of applicability." The central premise of special relativity is indeed a remarkably simple one: that the perceived speed of light never changes. But like those of any great prophet, Einstein's words were open to misinterpretation. Some of Einstein's conservative critics, not to mention the racist ones, conflated the theory of relativity with moral relativism. In truth it was closer to absolutism, for it set inviolable standards regarding how the universe behaves. Einstein later scolded Newton for relying on the "shadowy concept" of absolute space. The source of ultimate truth lies not in an invisible world, Einstein argued, but entirely within what we can perceive. Relativity grew out of Mach's philosophy, which states that we can know only what we can perceive, and in fact Mach had attempted to develop a relativity theory of his own. But in Einstein's hands physics also took on an element of theology. Newton placed ultimate reality in the hands of an unfathomable God. Einstein pulled back the curtain and linked reality directly to perception, making science the hotline to heaven.

Special relativity also proposed a new kind of unity, one between energy and matter. As far back as the ancient Greeks, philosophers had wondered about the relationship between an object and the impulse that causes it to move. In his *Optics,* Newton had speculated about the possibility of converting "bodies into light, and light into bodies." Einstein offered more than speculation. He explained how the conversion might work and gave an exact formula for the process, $E=mc^2$. The equation indicates that a minuscule amount of mass converts into a tremendous amount of energy—the source of power for deadly atomic bombs and for the life-giving radiance of the sun. It also shows that all forms of energy (heat, mechanical energy, and so on) are the same. When a burning log gives off heat, it looses a smidgen of mass in the form of energy; when a bowler sends a ball in motion down the lane, that kinetic energy slightly increases the ball's mass. Einstein's quantum physics created a related kind of unity by blurring the distinction between particles and waves.

At the same time, relativity theory dispensed with the nineteenth-century voodoo of the ether. To Einstein, the ether was as unnecessary and unknowable as its near twin, Newton's absolute space. "The introduction of a 'luminiferous ether' will prove to be superfluous inasmuch as the view here to be developed will not require an 'absolutely stationary space,'" he wrote in his 1905 paper. One of the guiding principles of relativity is that all observers see the same thing, so the nature of light is the same to all observers at all times. No wonder Michelson could not find any effect of light blowing in the ether wind. The new theory also accounted for other unexplained physics phenomena, such as the way that the mass of the electron seemed to increase when it was accelerated to an extremely high velocity. One consequence of the equivalence of mass and energy is that an object appears weightier the faster it moves—again, the example of the bowling ball—because the energy of motion also expresses itself as mass. An interstellar traveler trying to accelerate his starship to the speed of light would find himself endlessly frustrated. Getting a little closer to the magic velocity would take a lot of energy, which would make his ship more massive, so accelerating it further would take still more energy and so on. Getting all the way to the speed of light would take an infinite amount of energy and so is impossible.

Einstein braced for an explosive reaction to the publication of this theory but was promptly disappointed. What happened the day after Jesus delivered his Sermon on the Mount? Not much, probably. So it was with Einstein: calm followed his storm. "His publication was followed by an icy silence," his sister Maja recalled. Slowly, however, scientific recognition caught up with him. The prominent German physicist Max Planck took an interest in the paper and helped publicize its ideas. In 1908, Einstein finally left the patent office behind and moved to the University of Bern, where, with some string pulling by his allies, he began expanding the "area of applicability" of his special theory. He received some assistance in this regard from Hermann Minkowski, a Russian Jew living in Switzerland. A far more skilled mathematician than Einstein, Minkowski reconfigured the terms of the relativity equations to combine the three dimensions of space with time so that events could be described as occurring in a four-dimensional space-time, yet another novel scientific unity.

Interweaving time and space in this way set the foundation for the concept of curved space, which became a central part of Einstein's ideas about the cause of gravity and the structure of the universe. It also helped establish the notion that space and time cannot exist without each other and so marked yet another leap in the potential scope of scientific inquiry. Minkowski's version of space-time was a big step toward general relativity, but it was not enough in itself. The Minkowski equations described a flat geometry, like the Euclidian geometry that students learn in high school. Einstein needed a curved geometry to describe gravity, and this was his ultimate goal. Minkowski's thoughtful elaborations of relativity were cut short before he was able to help in this development. He died from pneumonia in 1909.

Six months later, Einstein formally joined the world of academia by accepting a position at the University of Zurich. He moved again, to Karl-Ferdinand University, the German university in Prague, just two years afterward. During his time there, he returned to his thoughts on the nature of gravity, this time considering its effect on light. Light is so intangible that we don't normally think of it as having weight, but Einstein believed otherwise. Starlight passing very close to the edge of the sun would be bent by the powerful gravity, displacing the star's apparent position in the sky away from the sun. Einstein realized this effect should be small but measurable. He had discussed this possibility before but examined it in greater detail, for a very practical reason: "I now see that one of the most important consequences of my former treatment is capable of being tested experimentally." In other words, he saw a chance to prove his still-controversial theory of relativity. "It would be a most desirable thing if astronomers would take up the question here raised," he concluded in the journal *Annalen der Physik* in 1911.

By 1912, Einstein had returned to Zurich, this time to take a position with the old Polytechnic. As his ideas about gravity and space continued to percolate, Einstein called on the mathematical talents of Marcel Grossmann, a friend since his student days. This beginning of general relativity brought Einstein another dose of agony. "Compared to this problem the original relativity theory is child's play," he wrote in a letter to Arnold Sommerfeld, the German physicist. He and Grossmann struggled for nine months and finally published a paper in 1913 that outlined a new theory of gravitation, but Einstein was not satisfied with the result and hesitantly titled the paper a "draft." The derived equations did not have the unique simplicity he craved, and they did not fit his thinking about the incorruptibility of causality. He was groping toward a theory of the most basic aspects of reality, and he had to make sure the result felt exactly right.

Always fond of thought problems, Einstein crafted a particularly fine one to illustrate the goal of his general theory of relativity. A man finds himself in a windowless room. He does not know where he is, but he notices that he feels his normal body weight bearing down on his feet. Is he in a room sitting on the surface of the Earth, or is he in outer space, atop a rocket that is accelerating the room, giving rise to a downward that only simulates gravity? According to Einstein, there is no way for the man to know, even if he is equipped with the most sophisticated scientific tools. Continuing with his Machist approach, he took what looked like an unfortunate subjectivity and found in it another scientific absolute. The reason the man cannot tell the difference between acceleration and gravity is that there is no difference; gravity and acceleration are perfectly equivalent. Making such an assertion is one thing. Deriving a mathematically rigorous theory that explains why the two are equivalent is, as Einstein painfully discovered, another thing entirely. It took him two more years to come up with a fully satisfactory answer.

Meanwhile, Einstein held out for observational confirmation that he was on the right track. If the sun's gravity really could bend starlight, the only time the effect might be visible is during a solar eclipse, when the moon blots out the sun's blinding brilliance so that stars lying right beside the sun's disk would be visible. Erwin Findlay-Freundlich, an assistant at the Berlin University Observatory, heard about Einstein's prediction and planned to gather the desired data during a solar eclipse taking place in Russia in August of 1914. Einstein scrounged for funding to support the trip and, after writing to George Ellery Hale of Mount Wilson Observatory, it looked as though he had succeeded in setting up an astronomical expedition. If the result was positive, Einstein wrote to Mach in 1913, "your inspired investigations into the foundations of mechanics . . . will receive splendid confirmation." Instead the guns of August touched off World War I and made the experiment politically impossible. In retrospect, however, the delay was a godsend for Einstein. His early calculation of how starlight would bend around the edge of the sun gave a number that was only half as large as the correct value he derived three years later. If the 1914 eclipse expedition had succeeded, it would have appeared to discredit relativity.

By the time war struck, Einstein had returned to Germany. He had accepted a post at the University of Berlin in 1913, in a deal that was sweetened with membership in the prestigious Prussian Academy of Sciences and the promise that he would not have to expend his energies lecturing. His newly elevated circumstances, along with his conviction that he had resolved the earlier mathematical problems with his theory, contributed to a growing sense of self-confidence that bordered on intellectual arrogance. Day by day he grew less concerned about the importance of observational confirmation. He had never been much of an experimentalist, but his devotion to Mach's ideas had led him to praise the primacy of observation. Increasingly, however, Einstein seemed dazzled by the scope and logical economy of his theories. At times he sounded convinced that he was on the path to decoding the fundamental nature of God, so small details of empirical confirmation no longer mattered much. "Now I am completely satisfied and I no longer doubt the correctness of the whole system, whether the observation of the eclipse succeeds or not. The sense of the thing is too evident," he wrote to Michele Besso. He had taken the leap of faith toward sci/religion.

Even as Einstein was retreating from observation toward intuition, he was getting a harsh empirical lesson on realpolitik. Soon after the outbreak of the Great War, ninety-three prominent German intellectuals—including Planck, Einstein's early champion and one of the founders of quantum theory—endorsed the "Manifesto to the Civilized World." In wounded terms, this document justified Germany's aggression as a necessary act of self-defense. In response, Einstein lent his name to the "Manifesto to Europeans," a denunciation of the war and call for European unity. The experience confirmed Einstein's deep antipathy toward what he saw as the aggressive and closed-minded nature of the German spirit and intensified his drive to find in physics the rational and unifying rules so grossly missing in the world of human affairs.

War paralyzed Europe for four years, while Einstein was deeply engaged in his own battle, his struggle for enlightenment regarding the nature of gravity. Newton himself had been disappointed by the descriptive theory of his theory of universal gravitation: "The Cause of Gravity is what I do not pretend to know," he wrote sullenly. For Einstein too it was not enough to know how that apple falls from the tree. He wanted science to penetrate more of the *why* of the universe. He became convinced that gravity results not from Newton's instantaneous, spooky force reaching through space, but through the interaction between matter and space itself. Notwithstanding his boasts to Besso, Einstein had great difficulty tying up the loose ends of his general theory of relativity. "Compared with this problem, the original relativity is child's play," he wrote to Sommerfeld in the autumn of 1915.

Einstein presented the results of his labor in a pair of lectures to the Prussian Academy that November. A full account appeared in the spring of 1916 as "The Foundation of the General Theory of Relativity" in *Annalen der Physik.* Newton's gravity had vanished, replaced by a field, springing from all matter, that bends space and any light or other radiation that happens to pass through it. A carpenter looks down a board to see if it has been planed cleanly. A child points at his ball to locate it. They take for granted that light travels in perfectly straight lines, and for our usual purposes it does. But Einstein's relentless logic led him to conclude that those lines are not perfectly straight. The geometry of space is never truly flat, according to general relativity, and the shortest path between two objects is never exactly a straight line. In the presence of very dense masses or over very long distances, the curvature of space becomes so significant that space bends, twists, even circles in on itself.

Or so Einstein claimed. In 1916, general relativity was a grand idea with no unique observational support. It drew its strength purely from Einstein's conviction that the rules and concepts of special relativity must apply to accelerating bodies, such as the hypothetical man in the sealed room, and that this expanded theory provided a more satisfying explanation of gravity than Newton's bewildered shrug. The obvious test, still, was to look for the curvature of light during a solar eclipse. Using his revised mathematics, Einstein now calculated that starlight passing by the edge of the sun would be deflected by less than two seconds of arc, an angle about one-thousandth as wide as the full moon. Finding such an effect would require extremely careful measurements of stellar positions. Once World War I was over the political obstacles were removed, but the test would have to wait for an eclipse that occurred in a part of the world where astronomers could make good observations and at a time when the sun was in a usefully starry location in the sky.

Fortunately nature offered another test, the weirdly migrating orbit of Mercury. Much as earlier discrepancies in planetary motions had led Kepler to abandon perfect circles, Mercury's unexplained motions hinted at a flaw in Newton's description of the force that generates the orbital ellipses. No matter how scientists tallied up all the gravitational influences in the solar system, they could not fully explain that spirographlike orbit of Mercury. Some of the greatest minds had tried and failed to find the fix that would patch up Newton's theory of gravity. A half century earlier, the French astronomer Urbain Leverrier had proposed that Mercury's precession was caused by the gravitational pull of an undiscovered planet, Vulcan, orbiting between Mercury and the sun. Leverrier had spectacularly predicted the existence of the planet Neptune based on the similarly errant motion of Uranus. This time around, however, dogged searches for the hypothetical Vulcan proved futile. In 1895, Simon Newcomb, one of America's leading astronomers, had even proposed tinkering with Newton's gravitational equations just a little in order to make the problem go away.

Einstein was well aware that nobody had found a satisfactory explanation for the precession of Mercury's orbit. The reason they all failed, he believed, was that they were all using Newton's theory of gravity, which worked well most of the time but failed under extreme conditions. Of all the planets, Mercury orbits closest to the sun's great mass and travels the most rapidly in its orbit. It therefore is the most strongly affected by the way the effects of general relativity warp the fabric of space and alter the flow of time. Preliminary versions of general relativity did a little better than Newtonian gravity at predicting the orbit of Mercury but still did not yield the exactly correct motion. Einstein scrapped these early drafts as mathematically flawed. But the final 1915 version of relativity not only satisfied Einstein aesthetically, it also got the orbit of Mercury exactly right, as he proudly informed the Prussian Academy of Sciences. Only then did he confess—and only in private—that he had kept throwing out his equations and devising new versions until they fit the data. As with his remarks about the solar eclipse experiment, Einstein's description of his efforts to model the motion of Mercury correctly show him growing increasingly wrapped up in his scientific theology. "I was only concerned with putting the answer into a lucid form. . . . There was no sense in getting excited about what was self-evident," he told journalist Alexander Moszkowski.

More and more, Einstein relocated his search for truth to the realm of pure thought. Nevertheless, he well understood the importance of observational support for his radical new theory. Perhaps he knew that his explanation of Mercury's orbit, no matter how tidy, would not carry the level of authority he needed. Other scientists claimed to have cracked this problem before. To convince the skeptics who did not see general relativity as "self-evident," Einstein sought better proof. A scientific prophet's claim becomes credible only when his predictions come true, and with general relativity Einstein was audaciously claiming mastery over all of space and time. He needed that eclipse test, much as the Israelites needed to see Moses' miracles or the Copernican view of the solar system needed Galileo's telescopic evidence to prove that the Earth circles the sun. But war made an eclipse expedition impossible. "Only the intrigues of miserable people prevent the execution of this last, new, important test of the theory," an exasperated Einstein wrote in another letter to Sommerfeld in 1915.

Einstein got his wish, although it took four more years and a roundabout set of circumstances. He had sent out a number of copies of his "General Theory" paper. One of these went to Willem de Sitter, an astronomer at the University of Leiden who would later play an important role in honing Einstein's cosmological notions. De Sitter in turn forwarded the paper to the eminent British astronomer Arthur Stanley Eddington, who immediately grasped the import of the work and, with the zeal of a convert, set out to prove it correct. Nature was on Einstein and Eddington's side. Astronomers noted that a total solar eclipse in 1919 would take place while the sun was nestled in the head of the constellation Taurus, a patch of sky inhabited by a cluster of stars called the Hyades. The region around the eclipsed sun would be packed with relatively bright stars—exactly the setting needed to make a workable search for the bending of starlight by gravity. At Eddington's urging, Sir Frank Dyson, Britain's astronomer royal, began preparing for an expedition to view the eclipse from the coast of West Africa. The observations took place as scheduled on May 29.

After much delicate analysis, Eddington announced his results to a joint meeting of the Royal Society and the Royal Astronomical

Society on November 6. The observation decisively confirmed Einstein's predicted bending of light. "The setting . . . resembled a Congregation of Rites," wrote physicist Abraham Pais, Einstein's colleague and biographer. The society's main hall overflowed with eminences ranging from the philosopher Alfred North Whitehead to J. J. Thomson, the man who discovered the electron. A portrait of Newton looked down upon the attendees. Cecilia Payne-Gaposchkin, a British astronomer and physicist who was still a student at the time, later recalled the impact of the event: "The result was a complete transformation of my world picture. . . . My world had been so shaken that I experienced something very like a nervous breakdown." The news media smelled the excitement. REVOLUTION IN SCIENCE, screamed a headline in next day's *Times* of London. "On November 7, 1919, the Einstein legend began," Pais said. Two days later, the story broke in *The New York Times.* The old religions were in trouble, even if they didn't know it. With Einstein's sudden celebrity, the authority of science was about to extend to times and places it had never gone before.

In truth, Eddington had not approached the observation with anything close to objectivity. He opened his heart honestly years afterward when he admitted, "We don't need an eclipse of the sun to ascertain whether a man is talking coherently or incoherently." Einstein's mystical faith in relativity was catching. Eddington believed in Einstein and discarded some of the eclipse photographs that he deemed "flawed" because they did not show the expected displacement of the stars. Nevertheless, Eddington's stamp of approval carried tremendous prestige, and the results were bolstered by a similar finding from another British team that had observed the 1919 eclipse from northwestern Brazil. Thus spake Einstein: He pronounced his theory confirmed even before Eddington announced his final results. Especially telling are his comments to a student at the University of Berlin, who asked him what he would have done if the observations had not supported general relativity.

"In that case I'd have to feel sorry for God. The theory is correct anyway," Einstein allegedly retorted. One imagines him pausing and unfolding his famously enigmatic smile, leaving the student wondering if Einstein was mocking God or merely mocking himself.

Einstein had grown increasingly assertive in elevating his instincts into a personal religious faith. He continued in the tradition of Newton, carrying out a divine project by revealing God's handiwork. But Einstein rushed down the path that Newton had only tentatively blazed, dictating what God should want or how God should have made the universe. Not surprisingly, Einstein didn't wait for the eclipse expedition or any other test before embarking on an even grander adventure. After publishing the completed version of his general theory of relativity, Einstein reset his sights and barreled ahead. He had just demolished Newton's theory of gravity. Now Einstein was out to demolish Newton's entire cosmology.

In his paper "Cosmological Considerations of the General Theory of Relativity," presented to the Prussian Academy of Sciences early in 1917, Einstein set forth a shocking new picture of the universe. At the time, the general theory had proven itself capable of explaining only the precession of Mercury's orbit—an impressive achievement, but far from definitive proof of the theory's validity. Undaunted, Einstein declared that he could derive the size and structure of the entire cosmos from his puny equations. His answer contradicted every theory that had come before. "Even for Einstein," writes biographer Ronald Clark, "this was playing for high stakes."

Once again, Einstein describes his intellectual work as a painful struggle, as if he were Jacob wrestling with an angel. The effort of writing this paper so taxed him that, he joked to his physicist friend Paul Ehrenfest, he felt he was facing "the danger of being confined to a madhouse." He was already mentally exhausted from the mind-bogglingly complex mathematical details of four-dimensional space-time. Why embark on another grand venture?

In part, Einstein had a specific scientific goal, based on his interpretation of Mach's ideas about the nature of inertia. Inertia is the tendency for an object to resist being moved; if you've ever tried rolling an automobile off the road, you have a good idea how much inertia there is in two tons of mass. But if there is no absolute space and absolute motion, how can there be absolute inertia? Mach's answer was that inertia "is nothing more or less than an abbreviated reference to the entire universe." Einstein had echoed this view as early as 1912. This linkage connected every little instance of inertia, such as the leaning of a car around a tight curve or the difficulty of getting a bowling ball quickly down the lane, to the distribution of distant stars. If Einstein took Mach's ideas seriously, his theory of general relativity would be incomplete until he derived the overall structure of the cosmos.

But another motivation behind Einstein's cosmological dabbling surely lay in his profound intellectual goals. He was interested in global solutions, physical theories so powerful and general that they could describe any and every location without producing any inconsistencies. Newton's unbounded universe struck him as irrational because it contained an infinite amount of mass scattered through an infinite space. In such an arrangement, every direction would lead to an object exerting a gravitational force, thereby producing an infinite gravitational field—a dynamical version of Olbers's paradox. All of these infinities would make the equations of relativity meaningless and the search for ultimate answers futile. When Newton recognized the problems with the infinite solution, he had called on God to bail him out. To Einstein, however, the physical laws and God were one and the same. Conflict between the two was not acceptable.

Then Einstein revisited Newton's other picture of the cosmos, in which we reside within a finite clump of stars surrounded by a ceaseless stretch of nothingness. To the astronomers of the time, who roughly knew the structure of our Milky Way but were still unsure about the existence of other galaxies, this model seemed plausible. Einstein rejected it, however, recognizing that such an island universe is unstable. Slowly but surely the stars would interact with each other and disperse, so that the island would eventually evaporate and become like grains of sand scattered across the endless sea of empty space. He also considered the island universe arrangement wasteful, because "the radiation emitted by the heavenly system of the universe will, in part, leave the Newtonian system of the universe, passing radially outwards, to become ineffective and lost in the infinite." He recoiled at the thought of such a "systematically impoverished" creation.

By the process of elimination, Einstein returned to the finite universe of his ancient predecessors. The hard-edged, spherical construction envisioned by Eudoxus and Aristotle was no longer scientifically acceptable, of course. In its place, Einstein sought a seeming paradox, a universe that has a limited size but no physical boundaries. He wanted a new creation, a universe unlike anything ever imagined before.

In this scientific article of faith, Einstein was indebted to Bernhard Riemann, a German mathematics wizard who had died young, his most innovative work largely unappreciated, thirteen years before Einstein was born. Riemann imagined warped geometries that follow very different rules from those of the familiar planar, Euclidian geometry. In his alternative worlds, parallel lines can meet and the angles of a triangle do not add up to 180 degrees, exactly as if somebody warped and twisted a page from a geometry textbook. Draw a big triangle on a basketball and you will see that the angles add up to more than what they would on a flat surface. Or draw the triangle on the inside of a bowl and all the corners look pinched in, with the totals of their angles equaling distinctly less than 180 degrees. Combining Riemann's geometry with concepts from his general theory of relativity, Einstein discovered he could construct exactly the kind of universe he sought.

According to relativity, matter warps the structure of spacetime so that it follows Reimann's rules. It is difficult to visualize what this means, but think about the bending of starlight as it passes by the sun. From the perspective of the light beam, it is not bending; it is following a linear path through the warp of spacetime. At first, Einstein had thought that only a huge mass would produce a significant effect of this kind. While he was contemplating how the eclipsed sun would displace the light of the nearby stars, he realized that the widely scattered stars in the universe would collectively distort the overall geometry of space. It is helpful to think of a two-dimensional equivalent: a large, thin rubber sheet. A weight placed on the sheet causes it to sag. If you keep adding more weights in different places, the whole sheet begins to dip and assume a concave shape. Likewise, all the stars, planets, and rocks in the cosmos bend the geometry of space, so it resembles the inside of that bowl. If there is enough matter, Einstein concluded, the bowl would close in on itself. "The curvature is variable in time and place, according to the distribution of matter, but we may roughly approximate it by means of a spherical shape," he wrote. In other words, the universe is a huge lumpy ball, finite but unbounded.

To be precise, the real universe would be not a ball but a four-dimensional sphere of curved, three-dimensional space. Such higher dimensions are pretty much impossible to visualize, ironic for a model allegedly grounded in Mach's philosophy that science should be rooted in experience. Einstein's cosmology was more like a divine revelation, a description of something so far beyond human scales and human comprehension that we can talk about it but never truly know it.

The simplest way to grasp Einstein's solution is to return to the rubber sheet analogy. He used a similar analogy in *Relativity,* a semipopular account of his newly expanded theory that he published at the end of 1916. Our sheet has curved in on itself so much that it is now an enclosed sphere, basically a balloon. Now imagine a two-dimensional being living on this three-dimensional balloon, actually embedded in the balloon's surface, who is trying to understand his world. Our poor two-dimensional friend—let's call him Trevor—cannot imagine a direction called "up" because he lives entirely within the balloon's surface. As Trevor explores his world, he never comes to an edge, yet if he travels long enough in one direction, he will complete a circumference of the sphere and return to this starting point.

Einstein, displaying a touch of false modesty, declared "with a moderate degree of certainty" that our universe resembles Trevor's, only in our case we inhabit three-dimensional space that follows a curved geometry in an imperceptible fourth dimension. (This type of language has become the standard rhetoric of sci/religion: make a whopper of a claim, then surround it with modest-sounding qualifiers.) If Einstein was correct, then an intrepid traveler in a speedy rocket could take off from the Earth, race away in an apparently straight line, and eventually return home without turning around, just like Trevor circumnavigating his balloon. Thus Einstein escaped the contradictions of Newton's infinite universe. And without returning to the absurdity of Aristotle's sharply drawn crystalline spheres, Einstein gave back a geometric sense of our place in the cosmos. It seemed inconceivable that a scientific theory could explain the entirety of an infinite universe. German physicist Max Born later gushed, "This suggestion of a finite, but unbounded space is one of the greatest ideas about the nature of the world which ever has been conceived." By making the universe finite, Einstein opened up the possibility that all of existence could lie within the grasp of human conception—and then he confidently started down that path.

This new conception of the universe bore little resemblance to what contemporary astronomers thought they were seeing through their telescopes. Back then, many of those researchers estimated the Milky Way was about fifty thousand light-years across and believed it was the only galaxy in the universe. The Milky Way alone seemed huge; recall that the sun's neighbor 61 Cygni is eleven light-years away, and even that distance is roughly seven hundred thousand times the span from the Earth to the sun. Einstein was envisioning a cosmos much, much larger still. His model also required that the matter be scattered more or less evenly through all of that copious space, not all clumped in one place. That smooth distribution, now known as the cosmological principle, meant that space had an overall uniform geometry and that the universe was dynamically stable. The cosmological principle was, in a way, an extension of Copernicus's idea that we do not live in a privileged position. He proposed that the Earth is just one of a group of planets. Einstein took that idea further and suggested that the properties of our part of the universe—including its density—largely resemble the properties of any other location. He thought this had to be true in order to build a universe that followed the rules of relativity and satisfied Mach's principle. How this beautiful model corresponded to the real universe was unclear at the time. "Whether, from the standpoint of present astronomical knowledge, it is tenable, will not here be discussed," he wrote, seeming to brush aside such concerns. He had a vision, and he believed in it.

But as with the prediction of curved starlight, Einstein wanted confirmation to back up his revelation. Because it was fully described by a set of equations relating the size of the universe to its density, Einstein's model of the universe suggested the possibility of plugging in numbers and restaging the ancient game of calculating the size of the outermost sphere of the heavens. "The exact number is a minor question," Einstein insisted. In private, however, he told his reporter friend Alexander Moszkowski he could estimate the universe is a staggering one hundred million light-years across. This was a hugely daring claim at a time when the basic distance scale of the universe was entirely unknown. Einstein notably did not include his size calculation in the published paper. By the time he discussed his cosmological model before the Prussian Academy of Sciences in 1921, he had backed away from this line of argument, recognizing that "the distribution of the visible stars is extremely irregular . . . so it seems impossible to estimate the average density." Still, this did not alter his essential point that the universe is finite and that, given the right information, he could derive its dimensions purely from his theory.

In developing his "Cosmological Considerations," Einstein had expressed no doubt about his conclusion that the universe had to be finite in size. The duration of the universe was a much more complicated matter, one that rapidly led Einstein to a paradox and to a fateful decision. His intuition told him that the universe must be eternal. But as Newton had learned centuries earlier, a universe that is finite in space tends not to be infinite in time. In the updated conception, the equations of general relativity implied that the curvature of space should change over time, producing either cosmic expansion or contraction. Einstein could not and would not renounce his beautiful theory, nor could he bring himself to renounce his classical belief in a static universe.

Faced with this conundrum, Einstein conceived his clever, if arbitrary, way to explain the situation. What if, he supposed, space produces a mysterious repulsion—"a mass-density of negative sign, distributed evenly through space"—that makes itself known only over very large distances? Then everything could remain in balance, and the theory of general relativity could live in harmony with the happy reality that the sky is not falling. So Einstein modified his gravitational equation and added a "universal constant," denoted by the Greek letter *lambda.* With this little mathematical trick, he created a universe that could stand still.

Lambda gave empty space an outward pressure whose strength

is proportional to distance. On small scales its effect would be too small to influence the well-studied motions of the planets, which would explain why nobody had observed it. Over large distances, however, its cumulative effect would negate the pull of gravity. Lambda was not based on any experiment or even on any physical theory. It existed only to keep the universe at rest without compromising general relativity. The invocation of Lambda demonstrated that Einstein was at last fully committed to his search for sci/religious transcendence. In his relentless drive for unity, he had tried to reconcile two incompatible ways of looking at the universe. In the short run his effort failed, but Lambda kept returning as spiritual aid for building a coherent model of the universe. When Einstein endowed space with structure, he gave authority to the intangible and thus helped to usher in the new sci/religious era. Gradually Lambda evolved a larger meaning, representing scientists' unshakable faith that a truly comprehensive cosmological model is possible, if only they can find the right X factor that will strip away the last veil of mystery and arrive at some kind of ultimate truth—what Einstein called "the secrets of the Old One."

Einstein knew that Lambda looked like an arbitrary embellishment of general relativity. In his 1917 paper he confessed, "In order to arrive at this consistent view, we admittedly had to introduce an extension of the field equations of gravitation which is not justified by our actual knowledge of gravitation . . . necessary only for the purpose of making a quasi-static distribution of matter." As for why the universe had to be quasi-static, he pointed to "the fact of the small velocities of the stars." Historians often take this explanation at face value. After all, Einstein was no astronomer. Leading scientists of the day believed that the universe was at rest, that the Milky Way was the only galaxy, and that the stars within it were not racing away from us, or toward us for that matter, at high speeds. If Einstein blundered, the common argument goes, it was not because he placed too little trust in observational data but because he placed too much. In other words, his faith had faltered.

But there was much more to Einstein's imaginative leap. In 1917, astronomers were in the midst of a heated debate about the nature of spiral nebulae, those wispy swirls of light that William Herschel had studied a century earlier. Modern analyses showed that some of the nebulae were moving away from us at hundreds of miles per second, far faster than any known stellar motions. Many, though not most, scientists believed these nebulae were in fact other galaxies like our Milky Way. In that case, the measured motions of nearby stars within the Milky Way would reveal nothing about the behavior of the universe as a whole. Einstein should have been receptive to these arguments. As a student he had been an avid reader of Kant, who pictured the Milky Way as just one of myriad island universes. And the existence of a multitude of galaxies scattered through space would in fact have bolstered his belief that matter must be evenly distributed in the universe over very large scales. So why did he ignore the provocative new astronomical findings while formulating his cosmology?

Perhaps bits of Einstein's conservative schooling had come back to haunt him. His writings drew freely from the history of philosophy, as when he expressed the general theory of relativity as a confirmation of Descartes's notion that there is no such thing as empty space. His deep-buried classical instincts told him that the universe could not have a beginning or an end. In this he may also have been guided not by Aristotle but by Baruch Spinoza, the seventeenth-century Jewish philosopher who described God as an impersonal, eternal force defined by natural law. Many philosophers, theologians, and scientists had imagined that the cosmos might be finite in scale, but even the most devoted biblical chronologists did not believe that God sprang into existence just six thousand years ago. Even if God existed outside of conventional space and time, as Saint Augustine proposed, God's eternal essence had to exist somewhere from which He could create the universe. As Stephen Hawking notes, the basic tools for describing an expanding (or contracting) universe have existed for hundreds of years. "This behavior of the universe could have been predicted from Newton's theory of gravity at any time in the nineteenth, the eighteenth, or even the late seventeenth centuries," he writes. But the cult of stasis held sway and may have drawn Einstein under its seductive spell.

Perhaps Einstein fell victim to one of his bouts of arrogant detachment, in which his dedication to the pursuit of pure thought made it seem unnecessary to consult with experts in related fields. He expressed this attitude with shocking frankness in a 1906 paper: "In view of the fact that the questions under consideration are treated here from a new point of view, I believed I could dispense with a literature review which would be very troublesome for me."

Or perhaps Einstein felt that providing a scientific framework for the construction of the universe was all that really mattered. Maybe the overreaching that inspired him to stick Lambda in the equations was, at the time, at least, a reasoned strategy that let him show the world such a cosmic model was possible. Einstein was growing increasingly confident in the beauty and simplicity of his theories, regardless of whether the supporting data yet existed. The 1917 cosmology paper marked an important turning point in that regard. Up to then he had focused on theories to explain specific phenomena—the photoelectric effect, the propagation of light, the nature of gravity. Afterward he shifted his attention to broader, unifying themes—first his cosmology, then his search for a theory that would show the underlying similarity of gravity, electromagnetism, and the two forces governing the behavior of atomic nuclei. And around this time he started speaking much more expressively about the romantic and religious underpinnings of his scientific work. "The supreme task of the physicist is to arrive at those universal elementary laws from which the cosmos can be built up by pure deduction. . . . The state of mind which enables a man to do work of this kind is akin to that of the religious worshiper or the lover," he said at a 1918 scientific gathering in celebration of Max Planck's sixtieth birthday.

The true answer surely involves elements of all three explanations. The last one is certainly the one that has had the most powerful repercussions. It also hints at a tragic element in Einstein's personality. His universe extended far beyond anything yet measured by astronomers, so he could easily and correctly have argued that local measurements reveal nothing about the cosmos as a whole. His belief in a uniform distribution of matter, for instance, already contradicted the scientific understanding of the day. Einstein could have argued equally forcefully that the equations of general relativity required a dynamic universe. Had he done so, it would surely be remembered as one of his greatest insights. But the pride that served Einstein well regarding other insights, such as the bending of light, led him astray this time. He was so sure he knew how God would have created the universe that he was blinded by his own faith.

For all its shortcomings, "On the Cosmological Considerations on the General Theory of Relativity" was a watershed paper that established Einstein as the true prophet of sci/religion. In it, he redrew the universe as utterly as Sigmund Freud had redrawn the map of the mind. He announced that science could explain the entire order of the physical world, much as Charles Darwin attempted to explain our place in the living world. Old-time religion had to give up its claims on the dark, infinite reaches of the cosmos. Einstein's cosmology was a manifesto for a new religion, one built explicitly on the joys of exploring God by exploring reality. "I am of the opinion that all the finer speculations in the realm of science spring from a deep religious feeling, and that without such feeling they would not be fruitful," Einstein explained some years later. But he made it clear that this religious feeling springs from the human intellect. "He believed in the power of reason to guess the laws according to which God has built the world," said Max Born, a physicist and close friend.

The casual ease with which Einstein created Lambda demonstrates just how ready he was to find the Lord in a set of equations. Although he soon had second thoughts about his cosmological constant, it was not because he had abandoned this spiritual path toward truth. In 1919 he distanced himself from Lambda, but only because he found it "gravely detrimental to the formal beauty of the theory." Years later, when he reportedly denounced Lambda as his "greatest blunder," it was because Lambda did not work as intended and improved observational evidence seemed to contradict it. Lambda was cast out of heaven, but the cosmology that replaced it was just as speculative and, in its own way, just as mystical.

Even then Lambda lived on. In the eight decades that followed, countless other researchers have followed Einstein's lead and invoked Lambda to produce a more aesthetically appealing universe or to tinker with models that don't seem to follow the observational data. Einstein's faith and yearning for scientific ecstasy were contagious. His example emboldened many other theorists to treat questions of origin, fate, and even theology as valid territory for science, expanding and intensifying the faith of sci/religion. And his ideas gave tremendous new weight to the studies of the spiral nebulae, those still-ambiguous objects whose true nature would soon bring God's craftsmanship into much sharper view.

CHAPTER FOUR

THE NEW CARDINALS BICKER IN EUROPE AND AMERICA

AFTER THE GREAT WAR THAT tore apart Europe, a higher-minded battle broke out in the world of cosmology. Actually it was two battles, inspired by similar spiritual yearnings but carried out in different fields. On the theoretical side, Einstein's finite but static model came under fire, as other physicists began to test the global implications of general relativity. Meanwhile, equally fierce clashes erupted among those who sought mystical truth through the eyepiece, not the equation, as astronomers debated whether our galaxy is unique or just one among a multitude. The mere existence of conflict was a sign of how far the authority of sci/religion had grown. Cosmology could now sustain its own version of the Reformation. Now it was possible to use mathematics to argue about the form and fate of the universe; now it was possible to place a yardstick to the heavens and disagree about the resulting measurements. The great book of nature lay open for the whole congregation to see. Everybody was eager to debate its passages, but nobody could agree on how to interpret their meaning.

Einstein had started the theoretical action with his 1917 cosmology paper and his self-critical comments about Lambda, which together invited other researchers to try their hand at modeling the universe. Then he promptly went missing in action, as the stresses tearing at him—trying to do science in wartime, dissolving his marriage to Mileva, and completing general relativity—became too much to bear. In 1918 he suffered a mental breakdown and experienced chronic stomach troubles that briefly convinced him he had cancer. During this time Elsa Lowenthal, Einstein's cousin, aided him in his recovery and developed an increasingly intimate bond with him. She became his second wife after he obtained a divorce from Mileva the following year. Even after he resumed his work, Einstein was preoccupied with more tangible and testable aspects of general relativity and largely turned away from cosmology for a few years.

While Einstein's attentions had drifted elsewhere, some of the more daring apostles of sci/religion revisited the equations of relativity, uncovering the limitations of his cosmological model. These critiques came from unlikely sources: Willem de Sitter, Einstein's friend and collaborator at the University of Leiden; Alexander Friedmann, an obscure Russian physicist with a knack for meteorology; and Georges Lemaître, a Belgian abbé with a flair for engineering. Almost as implausible was the nature of their criticism. It was not that Einstein had gone too far, but that he had not gone far enough in exploring the ways a small set of equations could describe the destiny of the universe.

De Sitter, a genial Dutch researcher with a firm grounding in both math and astronomy, followed a painstaking approach to research that alternately complemented and clashed with Einstein's broader, more intuitive way of thinking. As a student at University of Groningen, de Sitter had studied mathematics but found his true calling in the observational world. In 1908 he joined the Leiden observatory; he rose to the post of director and remained there until his death in 1934. At the observatory he excelled at unglamorous, highly detailed studies, such as determining the precise rotational motion of the Earth. With his tidy white goatee and eyes often lost in thought, de Sitter was the picture of the absent-minded professor—and in fact he was notorious for his good-natured forgetfulness. Far more than Einstein, de Sitter had the background and temperament to connect grand, cosmological ideas to the messy reality of the visible stars.

Ever since the introduction of special relativity in 1905, de Sitter had been fascinated by Einstein's new physics and especially by the possibility of discerning its observable effects. Around 1913 de Sitter studied double stars to determine if the speed of light from these orbiting bodies is truly independent of their motion; when he heard about general relativity in 1916, he energetically began to calculate the astronomical implications of the theory, starting by examining how it applies to the orbits of the planets. De Sitter was already at work on an English-language popularization of general relativity when Einstein visited Leiden late in 1916. The two men met and began a fruitful collaboration, sometimes more like a competition, that drew de Sitter ever deeper into Einstein's all-encompassing worldview.

Despite his innately practical and empirical outlook, de Sitter found himself inexorably drawn to the miraculous side of general relativity. Suddenly the universe was a plaything. Who could resist? While analyzing Einstein's equations, de Sitter found that they could be reformulated in ways that Einstein had not considered. In a trio of dense papers presented to the Royal Astronomical Society in 1917, de Sitter set out three slightly different interpretations of this discovery. In the final one, he hit on a strange but useful simplification and set the density equal to zero. This "de Sitter universe" maintained key elements of Einstein's universe—it was still uniform and it still had a tacked-on Lambda—but it contained no matter. De Sitter argued this approximation was reasonable because the density of the real universe is quite low. He called the resulting cosmological model "solution B" in deference to Einstein's original "solution A." The single voice of spiritual authority was sundered in two, like the split between the Essenes and the Pharisees in pre-Christian Palestine.

To Einstein, "solution B" was a serious blow. First off, de Sitter had shown that there was more than one way to interpret the cosmic implications of general relativity. Even worse, the alternate interpretation undermined some of the philosophical underpinnings of the 1917 cosmology paper. One of the primary reasons Einstein had conjured up a finite universe was to establish a reference frame of matter against which to measure inertia, but "solution B" seemed to imply that inertia could exist in a completely empty universe. Einstein had assumed that Lambda would guarantee there could be no solutions for the case in which the cosmic density is zero. He was sorely disappointed to find his equations had let him down. In a letter to de Sitter, Einstein complained that the empty universe "does not correspond to any physical possibility." But now he clearly knew he had not attained the one and only description of reality, what he called "the true state of affairs." The prophecy of 1917 remained to be fulfilled.

But "solution B" was significant in another way: it seemed to produce an observable effect that could be detected and verified. In the de Sitter universe, the curvature of space-time seemed to be continuously decreasing, like the inside of an inflating balloon whose surface is continually expanding and hence growing flatter at any one spot. Because of this effect, de Sitter discovered, a clock at a great distance would appear to run more slowly than a nearby one. Consequently, a beam of light passing between the two objects would stretch in time and grow redder, with the magnitude of the effect proportional to distance. "The lines in the spectra of very distant stars or nebulae must therefore be systematically displaced towards the red, giving rise to a spurious positive radial velocity," de Sitter wrote. As de Sitter knew, astronomers at the time were just starting to discover that many nebulae did indeed have shifts in their spectra that seemed to indicate rapid movement away from us.

This sounds like a description of an expanding universe, but not quite. Note that de Sitter described the velocity as "spurious." Following the lead of the great Einstein, he still regarded the universe as static and the reddening of light, then known as the "de Sitter effect," as an illusion. In spite of his adherence to the gospel of cosmic stasis, de Sitter had taken a decisive, if unintentional, step toward setting the universe in motion. After all, space did seem to expand in the de Sitter universe. The problem was that there was no place where an observer could stand and measure the changes; it was very hard to understand what "expansion" meant in the complete absence of matter. Arthur Eddington, the British champion of relativity and a master of clarifying complicated scientific concepts, attempted to cut through the confusion by means of a simple thought experiment. He placed two particles—too small to disrupt the overall assumption of zero density—in the de Sitter universe and allowed them to move in accordance with the equations. The two particles accelerated away from each other, as if the unraveling of space were a true, physical expansion. "A number of particles initially at rest will tend to scatter," Eddington declared. Other researchers disputed his interpretation. De Sitter, who had intended his cosmology to be static, was unsure what to think of this turn of events.

Few scientists could make heads or tails of de Sitter's ideas. Einstein, who agreed with Descartes that space existed only in relation to objects, did not think much of a cosmological model that allowed space but not matter. He also objected that the de Sitter effect led to an absurd result. Somebody sitting in the de Sitter universe would see time growing progressively slower the farther he looked. At some great distance, the slowing would become so extreme that time would appear to stand still, creating a kind of edge of reality where the world appeared trapped in a single moment. This was most certainly not the kind of unbounded universe Einstein's God would have created. Astronomers, meanwhile, were mostly baffled by the high-minded debate over real versus imaginary motions in this strange something called "solution B."

In many ways, de Sitter ended up raising more questions than he answered. But by applying his stubborn curiosity to the equations of general relativity, he helped excise some of the dreamy quality from Einstein's cosmological model. If there was a second solution to the equations of general relativity, Einstein realized glumly, there might be many. Perhaps the one that described the "true state of affairs" could be found among the others. He was never entirely comfortable with the arbitrary nature of Lambda but invoked it to keep the universe eternally at rest. Now de Sitter showed that Lambda didn't even do its assigned task. The universe could appear dynamic—or at least quasi-dynamic in some funny way—even with the cosmological constant. In his final paper, de Sitter explored the possibility of jettisoning Lambda, which he disdained because he felt it "detracts from the symmetry and elegance of Einstein's original theory." By exposing possible observable effects of a cosmology built around general relativity, de Sitter reminded Einstein that he, like Copernicus before him, needed hard evidence to support his revolutionary view of the universe. Good sermons alone don't get at God's secrets; scientific prophecy proves its power only when witnesses can testify to its truth.

Coaxed along by these thoughts, Einstein began the slow process of disowning Lambda. More than a decade later, he joined with de Sitter to deal with this problem and try once again to fashion a single, conceptually beautiful model of the universe. By then astronomers had reported shocking news that forced Einstein to abandon completely his ideal of a static universe. Until then, however, neither he nor de Sitter was prepared to confront the full physical implications of a universe ruled by general relativity. That monumental task fell to Alexander Friedmann, a largely forgotten visionary who was the leading disciple of sci/religion—the first person to break with thousands of years of tradition and set the universe in motion.

Friedmann's brief life was full of dramatic, improbable twists. He was born in St. Petersburg, Russia, to parents whose bent was artistic, not scientific. His mother was a pianist, his father a ballet dancer and composer. When Friedmann was nine years old, his parents divorced and he ended up in the custody of his father; he did not see his mother again for twenty-four years. Notwithstanding the domestic turmoil, Alexander shot to the top of his class at the St. Petersburg Gymnasium, where he was promptly caught up in another kind of unrest. The "Bloody Sunday" massacre of 1905 outside the czar's palace triggered widespread protests and student uprisings, which Friedmann joined. By the end of the year, Czar Nicholas II created a reformist constitutional monarchy that lasted until the revolutions of 1917 that brought the Communists to power.

While Russia lapsed into relative calm, Friedmann studied math and physics at the University of St. Petersburg. There he worked under Paul Ehrenfest, the Austrian-born physicist and animated free thinker who later struck up a close friendship with Einstein. Ehrenfest introduced the young Friedmann to relativity, quantum theory, and other new ideas sweeping physics. While continuing his graduate education, Friedmann also lectured at the university's Mining Institute, which contributed to his unusual brew of expertise in math, physics, aeronautics, and meteorology.

Friedmann's academic progress again halted abruptly, this time because of World War I. He enlisted as an aviator, lectured on aeronautics to Russia's fledgling air force, and buried himself in math theory between bombing raids. "Sometimes I get sick of the war . . . and yet the spirit is still strong, and if I get used to studying here, then probably, by the end of the war, I will have finished my dissertation," he wrote to his mathematician friend Vladimir Steklov. Although his letters frequently display a similar effort to keep his spirits high during war, Friedmann emerged from the conflict depressed and in poor health. He returned to St. Petersburg University—now Petrograd University, reflecting the revolution that engulfed Russia in civil war and ended with the triumph of the Bolsheviks under V. I. Lenin. (For a brief time, Russia had a leader who was also interested in the latest science: Lenin's library in the Kremlin included two dozen books on relativity theory.) Amazingly, Friedmann persevered through the changes and managed to finish his master's and secure a number of academic appointments in Petrograd. One of his students at Petrograd University was George Gamow, who later developed the idea of the expanding universe into a detailed model of the parable of creation.

After the end of World War I, word of Einstein's new theory finally filtered into Soviet Russia and into the underfunded, often unheated halls of Petrograd University. Friedmann immediately immersed himself in a detailed study of general relativity. Given the endlessly difficult circumstances of his life, no wonder Friedmann was so drawn to cosmology. Imagine his mathematical mind soaring above the warfare and petty bureaucracies that had made his life miserable, as he wrote in 1922, "The surest and deepest way to study the geometry of the world and the structure of our universe with the help of Einstein theory consists in the application of this theory to the whole world and in the use of astronomical research." Archival photographs from those years show Friedmann as a thin, bespectacled man, his receding hairline weakly offset by a limp mustache. Behind his watery eyes and introspective expression, however, a grand panoply of possibilities was unfolding.

When he worked out this global application of general relativity, Friedmann discovered that the static picture of the universe made no sense; the nature of the cosmos is to expand or contract. He kept Lambda in the equations and, like Einstein—but unlike de Sitter—assumed a uniform distribution of matter. Yet still space erupted into motion. Friedmann set out his new solutions in two stunning papers, 1922's "On the Curvature of Space" and 1924's "On the Possibility of a World with Constant Negative Curvature," both published in the prominent German physics journal *Zeitschrift für Physik.* Where de Sitter polished the equations of general relativity and found a second, ambiguous way to interpret their cosmological meaning, Friedmann gave the equations a more determined rub and resoundingly unleashed the genie that Einstein had tried to keep corked in a bottle.

In Friedmann's hands, the corrected equations of relativity allowed a dazzling array of possible cosmologies, every single one of them alive with motion: expanding universes, contracting universes, even oscillating universes that grow and shrink as if following the exhalations and inhalations of a cosmic Creator. Each of his solutions corresponded to a different geometry of space, a particular warping of the four-dimensional analogue of our distorted rubber sheet. An oscillating universe would be concave, or bowl shaped, as in Einstein's initial cosmological mode. A contracting universe would also have this shape, but a depressingly limited life span. Expanding universes might be convex, shaped somewhat like a riding saddle, in which case they would expand forever, or flat, in which case the force of gravity would exactly balance the expansion.

Friedmann did more than undermine Einstein's belief in stability. He also showed that curved space did not necessarily imply a finite universe. A concave universe would fold in on itself; this shape corresponds to Einstein's unbounded cosmos of limited extent. But flat and convex universes could be infinite in dimension. As de Sitter's work hinted, Einstein had failed to appreciate how much lay untapped within his equations. All of a sudden the relativity equations were birthing not just one or two but myriad possible universes, from which the theorist could pluck the most attractive ones like a shopper selecting a natty new tie. And unlike de Sitter's vague interpretations of the uncurling, empty cosmos, Friedmann's solutions were not just academically interesting descriptions of space. They were genuinely plausible models of space plus matter that could be connected to the real world. Friedmann lacked the skills or knowledge to follow through on his desire to link his theoretical models to astronomical observations, but that advance came soon enough. Scientists still refer to his three basic geometries of space and argue over which one best fits the latest data like Catholics and Protestants debating Scripture in sixteenth-century Switzerland.

Of all these cosmic possibilities, Friedmann considered the oscillating universe the most appealing, for it suggests a potentially endless cycle of rebirth. It thus avoids the notion of a beginning of time, a seeming impossibility that vexed Einstein, not to mention Saint Augustine, Newton, and, in recent years, Stephen Hawking. One could imagine that each expansion allows time for stars, planets, and life to appear; then everything collapses down to a point, resetting the clock so that the universe can rebound, expand again, and begin another existence. The modest Friedmann lacked Einstein's propensity for broad theological pronouncements, but now he could hardly ignore the religious implications of his work. "This brings to mind what Hindu mythology has to say about cycles of existence, and it also becomes possible to speak about 'the creation of the world from nothing,'" Friedmann wrote in *The World as Space and Time,* his 1923 book summarizing his work.

How long such a cycle would last depends on the mass of the universe, a number that was completely unknown at the time. "All this should at present be considered as curious facts which cannot be reliably supported by the inadequate astronomical experimental material," he wrote. But Friedmann was sufficiently intrigued that he took a crack at estimating the age of the universe, just as Einstein couldn't keep himself from speculating about its size.

Friedmann even cited Einstein's unpublished values for the radius and density of the universe. If we live in an oscillating universe, Friedmann guessed that the expanding cycle would last "of the order of 10 billion years." Other scientists had attempted to measure the age of the Earth or the ages of the stars, but here was something utterly new. Friedmann, in his matter-of-fact manner, presumed that science could derive a time for the beginning of existence, treading onto the territory once held securely by the biblical chronologists.

Like de Sitter, Friedmann seems to have viewed his cosmological calculations more as mathematical idealizations than as descriptions of the real universe. As he put it, "The data available to us are completely inadequate for any kind of numerical estimates and for solving what kind of world our universe is." But the genie was out of the bottle. From the 1930s on, astronomers gained more knowledge about the overall density and dynamics of the universe and started to speak more literally about the size, age, and rate of expansion of the universe. Soon cosmic speculation knew no bounds.

The other two Friedmann solutions are less dramatic. The concave, or saddle-shaped, universe does not contain enough matter to pull itself back together, so it keeps expanding and never turns back. The third possibility is a flat universe, one in which the geometry is just like the Euclidian planes taught in primary school. Returning to the two-dimensional analogy, a flat universe is like a rubber sheet in which the local bumps and ripples average out so that the sheet as a whole resembles an even tabletop. Owing to the exact balance between space and matter, the expansion grows slower and slower, eternally approaching but never quite reaching a complete stop.

Friedmann never explicitly described a flat universe, though it is an implicit possibility in the range of cosmological models he portrayed in his two papers. This scenario struck Einstein as the one closest to the static universe he had initially preferred. When he finally had to abandon Lambda and pick one of the dynamic cosmologies, he flirted with Friedmann's oscillating universe but finally settled on a flat geometry, which he formalized in a 1932 paper co-written with de Sitter. Over time, a majority of cosmologists followed suit, and this "Einstein–de Sitter universe" became the leading contender. Modern theorists still consider a flat universe the most appealing version and sought support for it in their speculative equations long before astronomical observations showed it to be a plausible answer. Those new observations, ironically, are the ones that have resurrected the long-derided Lambda.

"Friedmann's papers laid the foundation for cosmology based on general relativity," reflects MIT's Alan Guth, who helped establish the modern incarnation of the flat universe. But the revelatory nature of Friedmann's work, so clear in retrospect, was not much appreciated at the time. After the publication of the 1922 paper, Einstein wrote a dismissive reply to *Zeitschrift für Physik.* "The results concerning the nonstationary world . . . appear to me suspicious. In reality it turns out that the solution given in it does not satisfy the field equations," he wrote. A curious use of the word *reality,* as the two men debated their mathematical idealizations of the universe! Einstein believed that Friedmann's debunking of the static universe arose out of a conceptual error and that, when corrected, "the significance of the work is precisely that it proves this constancy." This response struck a sour tone, suggesting Einstein's deeper objection was that Friedmann's dynamic universe did not sound the gong of divine truth.

In December of that year, Friedmann wrote to him to explain how he arrived at his conclusions but received no response. By the time the letter arrived, Einstein was off traveling in Japan. In the immediate aftermath of his international celebrity, he was suddenly deluged with more mail and callers than ever before, and he probably never saw Friedmann's note even after he returned. Einstein habitually retreated from distractions when immersed in his work. Earlier, while he was completing general relativity, he felt so overwhelmed with mail that he snared his letters on a meat hook hanging from the ceiling of his apartment. According to Einstein's friend Erwin Finlay-Freundlich, he burned the lot of them when the hook got full.

Friedmann persisted, attempting to visit Einstein in Berlin, but to no avail. His ideas might have vanished into oblivion but for the intervention of Yuri Krutkov, a colleague of Friedmann's from Petrograd University. Krutkov met Einstein in Leiden in the spring of 1923 and called his attention to Friedmann's work. Forced to reconsider, Einstein revisited the arguments of his 1917 paper and finally had to admit that the unknown Russian had defeated the famous German at his own game. "My criticism . . . was based on an error in calculations. I consider that Mr. Friedmann's results are correct and shed new light," he wrote in a follow-up letter to *Zeitschrift*. "It has turned out that the field equations allow not only static but dynamic . . . solutions for the space structure." Einstein still doubted that Friedmann's equations described that elusive thing known as the "real universe." But he was having serious second thoughts about Lambda, which he now considered "a complication of the theory, which seriously reduces its logical simplicity." Like Newton, Einstein tried to dictate how the universe should behave, only to find his own equations fighting him. It was becoming increasingly evident that Friedmann was correct: general relativity required the cosmos to move.

Friedmann was elated. "Everybody was very impressed by my struggle with Einstein and my eventual victory," he wrote. He was especially excited that he would now find it easier to get his papers published. Nobody will ever know how far Friedmann might have continued with his incisive cosmological analysis. In the summer of 1925 he set out on a scientific ballooning experiment designed to collect meteorological and medical data; along the way, he set an altitude record of twenty-four thousand feet. Perhaps the stress of the expedition was too much for him, because he died in the fall, just thirty-seven years old. The official diagnosis was typhoid fever, although Gamow believes Friedmann actually was done in by pneumonia contracted during his balloon flight. Despite his earlier response, Einstein did not follow up on Friedmann's ideas, perhaps because of his profound dislike of the concept that the universe could have emerged from a single point, the seeming implication of an expanding Friedmann universe. Friedmann's work remained little noticed in theoretical circles for nearly a decade, and utterly unknown to the astronomers who could have helped confirm his ideas. But the expanding universe soon resurfaced from an unexpected corner.

In the mid-1920s, a Belgian cleric named Georges Lemaître began looking into the global implications of relativity in much the same way Friedmann had several years earlier. Like his predecessor, Lemaître had fought in World War I—in this case, as an artillery officer in the Belgian army—before returning to the world of science and engineering. During the war he participated in fierce urban fighting and witnessed one of the first military attacks using poison gas. In other ways, Lemaître led a very different kind of life than did Friedmann. A gregarious, kind-faced man, he enjoyed a comparatively comfortable lifestyle and close contact with some of the major players in the nascent field of cosmology. He also sampled both the religious and scientific sides of life in a way that would have been unthinkable in Friedmann's postrevolutionary Russia. After the war Lemaître attended the University of Louvain but then entered the seminary and was ordained in 1923. Refusing to abandon his interest in mathematics and physics, he went on to study at the University of Cambridge and at the Massachusetts Institute of Technology. His studies persuaded him that the universe operates according to simple, knowable rules that may be obscured by complicated appearances. He wrote a comment in his notes that plainly expresses a sentiment both scientific and religious: "Behind objects that can be touched or looked upon, should be something hidden."

In England Lemaître studied under Arthur Eddington, the redoubtable champion of Einstein's relativity, who informed him about the apparent high velocities of certain spiral nebulae, which Eddington thought might be signs of the curious reddening of light predicted by de Sitter's cosmology. The following year, Lemaître continued his education in the United States, with a specific goal of learning more about "the astronomical consequences of the Principle of Relativity." While at the Massachusetts Institute of Technology, he took a trip down to Washington, D.C., and attended a meeting of the National Academy of Sciences. There, he heard the American astronomer Edwin Hubble lecture on his new discoveries revealing the scale of the universe. Lemaître also spent time at Harvard College Observatory studying under Harlow Shapley, another brilliant cosmic cartographer, and visited the Lowell Observatory and Mount Wilson Observatory, the leading centers for the study of the enigmatic spiral nebulae. These experiences bolstered Lemaître's faith that Einstein's equations could embody the hidden rules that would explain the whole of the universe.

In 1925, Lemaître returned to Brussels and took a position at the University of Louvain, where he wrote up a cosmological paper, "Note on de Sitter's Universe," that incorporated the lessons from his travels abroad. Most significant, he explored the "nonstatical" nature of de Sitter's cosmology and described it as "a possible interpretation of the main receding motions of spiral nebulae." Although he apparently had not read Friedmann's work, he then retraced it almost step by step and arrived at very similar results. Once again, he rejected Einstein's insistence on a static universe. Even with Lambda in place, Lemaître discovered (or rather, rediscovered) that there are whole families of solutions to the equations of relativity that do not produce a static universe. In fact, the static solution hangs in precarious balance; even a slight change in physical conditions could upset the equilibrium and cause the whole system to start expanding. He pictured a universe that expanded indefinitely outward from an initial state, which prompted him to ponder what happens if you run the clock backward and consider where this expansion began. He started to think that there must have been a time when all of existence was packed together in a compact, formless blob, which at some point came undone and begat the modern universe. Slowly but surely, Lemaître was edging past Friedmann and developing a full-fledged, physical description of a cosmos in motion.

These fresh ideas came just as astronomers were launching into their own Great War over the size and state of the universe. Such information was vital for evaluating the mind-bending cosmologies Einstein had formulated. When he developed his cosmic model in 1917, most scientists still believed that the Milky Way, our home galaxy, was unique and all-encompassing. The stars within the Milky Way follow leisurely orbits, and our galaxy is indeed quite stable, which allowed Einstein to envision an eternal, immobile universe—but was such a picture correct? On the other hand, Einstein's universe worked only if the shape of space-time folded in on itself. That required a high density of matter spread evenly throughout an enormous cosmos—but could the universe actually be like that? And there was that century-old puzzle, Olbers's paradox, hanging in the background. Could Einstein's finite universe explain why the night sky is dark?

By 1932, the consensus view had changed and many of the key questions were being answered. The fifteen-year period of scientific development after Einstein's cosmology paper was reminiscent of the upheaval of the early seventeenth century, when Galileo produced potent evidence in favor of the Copernican system and Kepler finally ended the cult of spherical motion. Theory fed on observation, observation fed on theory, and in the end science ended up much grander and more powerful than before. This time around, astronomers recognized Milky Way as one of countless galaxies strewn through a vast and dynamic universe, each one composed of many billions of stars. And the cosmologists were ready to make sense of it all, to demonstrate they could explain our place among the fleeing galaxies as readily as their ancestors had put the Earth in motion among the planets.

The path to enlightenment was achieved not just through better instruments but also through passionate philosophical debate. There is no need for a powerful telescope to see another galaxy. In fact, all it takes to spot a galaxy is good eyesight. Go outside on a crisp fall night and run your gaze along the bent narrow "V" of stars that marks the constellation Andromeda. Just above the arc you will perceive a wispy glow, a little smaller than the full moon. That glow is the Andromeda galaxy, a near twin of the Milky Way now estimated to lie about two million light-years away; it is the most distant object visible to the unaided human eye. A decent pair of binoculars will easily reveal a dozen other galaxies, if you know what you are looking for.

The difficulty in spotting other galaxies was one of conception more than perception. Astronomers simply could not accept the vastness of the universe, and the diminution of our place within it, that necessarily followed from accepting that the Milky Way is just one galaxy among a multitude. If anything, they had made backward progress since the eighteenth century, when William Herschel had begun cataloging the different types of luminous nebulae strewn across the sky. Herschel's argument that some of these nebulae were actually giant systems of stars, dimmed by their tremendous distance from us, carried considerable authority for a time. In the mid–nineteenth century, William Parsons, the third earl of Rosse, noticed that some of the nebulae had a lovely spiral form. A few decades later, the new technique of spectroscopy showed that the light from these spirals resembled that from stars, not from luminous gas.

By the beginning of the twentieth century, however, the tide of opinion had turned. The primary objection was geometric. Most of the spiral nebulae appear in parts of the sky far away from the band of the Milky Way. If these were external galaxies, the reasoning went, they should appear scattered evenly across the sky. The fact that they appear to flee from the Milky Way suggests they are secondary objects ejected from our galaxy. The real answer, discovered much later, is that the Milky Way contains light-absorbing gas and dust that blots out the light from any object lying behind it. A few researchers suspected as much at the time. But in those early days of the new sci/religion, many scientists still carried a subtle reactionary agenda in their hearts. In 1905, Agnes Clerke, a highly influential astronomer and scientific historian, wrote, "No competent thinker, with the whole of the available evidence before him, can now, it is safe to say, maintain any single nebula to be a star system of co-ordinate rank with the Milky Way." There is a whiff of the haughtiness of old religious cosmologies in these words: if our galaxy isn't the only one, at the very least it is by far the greatest. Sci/religion would soon establish the exact opposite notion as its approved doctrine. Until astronomers could determine the distance scale of the universe, they could make no connections between what they were seeing and the exciting but abstract cosmological ideas that Einstein had inspired.

In April 1920, the National Academy of Sciences in Washington, D.C., convened a debate (now recalled simply as the "Great Debate" because of its historical importance) to shed some light on the issue. Heber Curtis of the Lick Observatory in Santa Cruz, California, argued that the Milky Way was small in comparison to the vastness of space, and that the spiral nebulae were surely other galaxies lying at tremendous distances. Harlow Shapley from the Mount Wilson Observatory, just a couple of hundred miles south in Pasadena, took the opposing side. His view was that the Milky Way was about ten times as large as previously thought, so huge that it dominated the universe. Those vexing nebulae were just some gaseous junk spinning away from us, though Shapley understandably could offer only vague suggestions of how they got there or why they were in such a hurry to get away.

Curtis and Shapley argued their positions in styles as different as their viewpoints. Curtis was forty-eight, thirteen years older than Shapley, and with his round spectacles and dapper suit he looked rather patrician. He already had a well-established career at Lick, and he spoke with an orator's polish as he delved through a fairly technical presentation somewhat enlivened by a number of slides. Shapley, who resembled a rumpled Peter Lorre, took a more direct approach. As a former newspaperman, he knew better than to bludgeon his congregation with data. He delivered instead a directly spoken, broadly themed homily—although he, too, quickly introduced the many specific observational details he needed to press his case. Members of the National Academy of Sciences who wished to have a few drinks to ease them through the debate were out of luck. The Nineteenth Amendment had taken effect earlier that year, so the academy had to dispense with the wine it normally served before such lectures.

Historians of science have come to view the Great Debate as a crucial prelude to the dramatic astronomical discoveries that occurred later in the decade. Once scientists understood the size and nature of the spiral nebulae, they quickly understood a great deal about the size and structure of the universe writ large. For Shapley, too, this was a pivotal event: he was bucking for the directorship of Harvard College Observatory, knowing that one of the key members of the observatory's visiting committee was attending the debate. The actual contretemps, however, depended on a series of intricate arguments that sailed serenely over the heads of most of the audience, few of whom were astronomers. It garnered little attention in either the popular press or the scientific literature of the time. Sometimes great moments in history make themselves known only in retrospect. "I had forgotten about the whole thing long ago," Shapley wrote in his 1969 autobiography. "To have it come up suddenly as an issue, and as something historic, was a surprise." Eventually the Great Debate became known as pivotal confrontation of ideas, the Diet of Worms to the cosmological revolution that lay just a few years ahead.

At first glance, Curtis would seem the more progressive thinker of the two men. He maintained that "the spirals are not intragalactic objects but island universes, like our own galaxy, and that the spirals, as external galaxies, indicate to us a greater universe into which we may penetrate to distances of ten million to a hundred million light years." He was right and even a bit conservative. The deepest images from the Hubble Space Telescope show objects roughly thirteen *billion* light-years distant. In a series of clever back-and-forth comparisons, Curtis showed that the properties of the spiral nebulae seemed strange if they were located in our galaxy but made perfect sense if they lay in far reaches of space. He reiterated that the spectra of these nebulae—that is, the patterns of their light when passed through a prism—strongly resembled those of stars, not clouds of gas. He correctly guessed that obscuring matter in our galaxy explains why no spirals appear along the band of the Milky Way. And he cited studies of the enormous speeds of the spiral nebulae. "Their space velocity is one hundred times that of the galactic diffuse nebulae . . . such high speeds seem possible for individual galaxies," he said. In closing, he added a personal comment that shows how deeply mystical concerns had penetrated the world of observational astronomy: "There is a unity and an internal agreement in the features of the island universe theory which appeals very strongly to me."

But the debate was not as black and white as it appeared on the surface. While Curtis was largely correct in his views about the reality of other galaxies, he was in many other ways a scientific conservative. A year after the eclipse expedition that seemed to prove general relativity, Curtis remained skeptical of Einstein's radical theory and its cosmological implications. In the Great Debate, he severely underestimated the size of the Milky Way because he still placed his trust in star-counting measurements, an old-fashioned reckoning technique dating back to Herschel. Curtis distrusted Shapley's brand-new, and much more accurate, measurement techniques that relied on the study of certain types of variable stars.

Here we must have some sympathy for Curtis, however, because nature had played a nasty trick on him. In 1885 a bright star had flared up in the Andromeda, one of the brightest of the spiral nebulae. Assuming this was a nova, an erupting star similar to the ones seen in our galaxy, the Andromeda nebula couldn't be very far away. That in turn meant that it had to be fairly small. If nebulae were galaxies similar to our own, then the Milky Way had to be small as well. But Shapley's variable stars indicated the Milky Way is huge. Curtis was backed into a corner, so he had to argue against the way his opponent measured distances. What he didn't know—what nobody understood at that time—was that the 1885 star wasn't a nova. It was a supernova, a much brighter type of stellar explosion that is readily visible from much greater distances.

It is harder to forgive Curtis for his other bit of pigheadedness. Again following convention, he believed that the Earth lies almost at the exact center of the Milky Way. Such had been the party line ever since Herschel conducted a galactic census, tediously cataloging every star he could observe. In fact, we lie way out in the celestial suburbs, some two-thirds the way to the edge of the disk, but dusty gas clouds blot out much of the light from the dense concentration of stars at the center. The clouds create an illusion that the galaxy is equally bright in all directions, as if we were smack in the middle.

Shapley had found a way to part the interstellar mists and glimpse our true place in the Milky Way, using a technique that four years later definitely settled the entire dispute over the nature of the spiral nebulae. Here he drew on an amazing discovery by Henrietta Swan Leavitt, one of Harvard College Observatory's "computers"—the people who did the mundane but crucial work of surveying thousands upon thousands of stars and cataloging their properties. Starting around 1908, the meticulous Leavitt turned her attention to a unique class of stars called Cepheid variables, so called because the best-studied star of this type lies in the constellation Cepheus. These stars grow brighter and dimmer in a predictable, endlessly repeating manner. More extraordinary, she found that the period of variation correlates directly with the star's real, intrinsic luminosity: the longer the period, the more luminous the star. Ejnar Hertzsprung, a Danish astronomer who developed the still-standard scheme for classifying stars by color and luminosity, completed Leavitt's work by determining the approximate distances to a few nearby Cepheids, thereby anchoring the whole system of measurement. Now astronomers had reasonable hope that they could use the mathematical rigor of sci/religion to quantify the once-unfathomable depths of the cosmos.

Leavitt's intellectual process was roughly the inverse of Einstein's. He started with his visionary theory and worked his way down; she started with photographic plates from Harvard's twenty-four-inch refracting telescope and built up conclusions from what she saw. But humble sky analysts like Leavitt, a mostly female group laboring away like nuns in the cloister, were every bit as important as the high-profile cosmologists in extending the mystical power of modern science. Her research rested on the same kinds of optimistic assumptions as Einstein's 1917 cosmology. She believed in the constancy of physical law across time and space. She believed in the ultimate knowability of the universe. If Einstein was the prophet, Leavitt and her ilk were the quiet upholders of the faith who went out and witnessed science's miraculous power to bring the distant galaxies within reach. They spread the gospel by systematically attaching numbers to the different parts of the universe. Leavitt was a champion number cruncher.

Before she discovered the clocklike predictability of the Cepheids, astronomers were like mariners squinting at a light on the horizon. Not knowing its distance, they could not say for sure what they were looking at. It could be a mighty bonfire in a distant city, or it might be a puny lantern in the window of a seaside cottage right off the ship's bow. Leavitt found what scientists call a "standard candle"—a lighthouse beacon of known luminosity that allows an accurate plotting of the celestial seas and shoals. These beacons followed a remarkably regular pattern in which the period indicates the true luminosity. If all slowly flashing lighthouses had very powerful bulbs while the rapidly flashing ones had weak ones, you could estimate how far you are from shore just by timing the pulses of light. Similarly, a careful observer can determine a Cepheid's true luminosity and hence its distance just by plotting its changing brightness over several weeks. Cepheids are what makes the scale of the universe comprehensible. They were the instrument of Shapley's greatest triumph and later his greatest disgrace. Soon they unraveled Einstein's cosmology as well.

Shapley used the new Cepheid technique, along with the mighty eye of the sixty-inch reflecting telescope at Mount Wilson, to map the size and shape of our galaxy. He got the size of the Milky Way wrong—he thought it was about three hundred thousand light-years across, three times larger than the currently accepted number, due to a misreading of his variable stars—but he got the layout dead on. We live not at the center of the Milky Way, as everyone from Herschel on had believed, but far off to one side. Shapley cast himself as a modern Copernicus, weaning mankind from one more delusion about its central importance. "The significance of man and the earth in the sidereal scheme has dwindled with advancing knowledge of the physical world . . . we have reached an epoch, I believe, when another advance is necessary," he told his audience at the National Academy of Sciences. He implicitly accepted the modern view that our knowledge of the universe, not our place in the universe, is what makes us special. "Our conception of the galactic system must be enlarged to keep in proper relationship the objects our telescopes are finding; the solar system can no longer maintain a central location," Shapley said.

Curtis could not bring himself to accept these findings. Partly he questioned Shapley's measurements because they made the Milky Way improbably enormous, "at least ten times greater than formerly accepted." If our galaxy were that large, it would seemingly engulf the nearer spiral nebulae and hence cast doubt on the island universe interpretation that Curtis found so appealing. Such a drastic size revision demanded airtight proof, he insisted, and he distrusted the new and relatively unproven method of measuring distances with Cepheids. He questioned the basic assumption behind this technique, that the Cepheid variable stars behave the same way "anywhere in the universe." Curtis was a cautious researcher, the type who typically ended statements with the mantra "More data are required." But in his arguments there's also a hint of deeper skepticism regarding Shapley's optimistic belief in the uniformity of nature. Curtis just didn't have the faith.

In many ways it was Shapley, not Curtis, who was the more devout disciple of sci/religion in the Great Debate. Where Curtis kept calling for additional data, Shapley unflinchingly exploited up-to-the-minute techniques. Shapley's mistake was primarily one of recklessness. He put so much faith in his Cepheids and in the principle of uniformity that he didn't recognize the sources of error that caused him to overstate their distances. Based on those measurements, "the spiral nebulae can hardly be comparable galactic systems," he baldly claimed. They could scarcely compare to his puffed-up picture of the Milky Way. Possibly they were minor stellar groupings basking in the grand glory of our galaxy or, more likely, they were nothing more than the gaseous whirlpools they appeared to be.

At the very end of his lecture, Shapley added an interesting caveat. "There may be elsewhere in space stellar systems equal to or greater than ours—as yet unrecognized and possibly quite beyond the power of existing optical devices and present measuring scales," he said. Many astronomers at the time similarly believed that part of the universe lay beyond human perception. Einstein's 1917 cosmology, which boldly asserted that the rules of general relativity hold everywhere through a finite reality, inaugurated a shift away from this line of thinking. Even then, Lambda represented a kind of hedge in Einstein's nascent cosmic religion. To embrace the cosmos—truly everything—he needed an extra factor that was justified neither by theory nor by observation. But just a few years later, theory and observation began to merge, Lambda fell by the wayside, and cosmologists rapidly expanded the empire of numbers throughout all of space and time.

The Great Debate called attention to some of the novel astronomical techniques and observations that would make this possible. Yet one of the most remarkable discoveries of the day occupied only a minor role in the event. As Curtis had briefly mentioned, some of the spiral nebulae appeared to be moving at very high speeds, more than seven hundred miles per second. He took these tremendous velocities as evidence that these objects could not reside within our galaxy, where far more sedate motions are the norm. But there was much, much more to these peculiar motions, as soon became clear. Measurements of the anomalous motions of the spiral nebulae provided a key piece of the evidence that discredited Lambda, blew apart the age-old notion of the static universe, and ushered in the first creation story rooted in science.

In 1920, however, tracking the flittings of spiral nebulae seemed a tedious and not terribly exciting job. Basically the task fell to one man, a hardworking and little-known astronomer by the name of Vesto Melvin Slipher. He was one of those brilliant second-string players who lay the groundwork for somebody else to come along and grab all the glory—in this case, Edwin Hubble. There is no Slipher Space Telescope, there are no stirring biographies to inspire future generations of scientists. Yet Slipher's studies marked the essential first step toward the discovery that we live in an expanding universe. Had he worked with better instruments or possessed a more flamboyant personality, Slipher might be the one enshrouded in fame. Instead he is remembered, like Leavitt, as a diligent number cruncher and a historical footnote.

Slipher started down the road to near fame in 1901 when he left Indiana University to take on a "temporary" position at the notorious Lowell Observatory in Flagstaff, Arizona. He ended up staying for fifty-three years. Percival Lowell, a member of the aristocratic Lowell family of Boston, had built this lavish observatory as a shrine to his quixotic scientific fixations. Most famously, he believed that Mars was inhabited by an intelligent, dying race that had crisscrossed the planet with canals to distribute a dwindling supply of water. He drew wildly detailed maps of these canal systems, which he swore he could see through his telescope. The canals of Mars found their way into popular lore; Lowell's hyperbolic books became best-sellers. When NASA's modern spacecraft visited Mars, they spotted a giant canyon and various dark markings that Lowell might have seen and misinterpreted. The planet abounds with unique geological features, including giant volcanoes, ancient riverbeds, and windswept craters. There are, needless to say, no canals.

Despite the Mars fiasco, Lowell's cranky obsessions yielded some serious results regarding the pantheon of our solar system. His unbending belief in a planet beyond Neptune led to the serendipitous discovery of Pluto by Clyde Tombaugh, who was working under Slipher's direction. And Lowell's fascination with how planetary systems form inspired him to direct Slipher to study the spiral nebulae. Lowell was of the opinion that those nebulae were swirling clouds of gas caught in the act of condensing into stars and planets, a view also shared by a number of saner scientists of the day. He wanted to document this process and instructed Slipher to measure the presumed whirlpool-like motions of the nebulae. The meek Slipher and flamboyant Lowell made an odd pair, but they proved strangely suited to each other. Slipher's perpetually formal attire appealed to his boss's patrician tastes, and his measured approach to research counterbalanced Lowell's reckless, eccentric speculations. Lowell had such faith in his protégé that he named Slipher his successor as head of the observatory. After Lowell's death in 1916, the observatory and all its resources were at Slipher's disposal.

Starting around 1910, Slipher set out on the delicate task of measuring the motions of the spiral nebulae. He did so by looking at slight changes in the frequency of light caused by motion. Light waves from an approaching object pile up on top of one another, causing the waves to compress to slightly shorter, bluer colors. Light waves from a receding object behave in an opposite way: they stretch out, moving to the longer, redder end of the spectrum. Sound waves do the same thing, which is why the pitch of an ambulance siren or a car horn suddenly changes when the vehicle whips by, from a slightly higher (shorter wavelength) tone to a deeper (longer wavelength) one. This effect of motion on wavelength is known as a Doppler shift, in honor of the Czech mathematician Christian Johann Doppler, who in 1842 described the effect for sound and correctly predicted it would affect light as well. You needn't wait for an ambulance to pass to experience the phenomenon. You can hear the Doppler shift in action by warbling your voice in the comfort of your own home if you sing a steady pitch into the spinning blades of a box fan.

By the time Slipher set to work, examining Doppler shifts had become a well-established astronomical technique. The difficulty lay in gathering enough light from the nebulae, whose brightness is spread out over a sizable patch of sky rather than concentrated in one spot like a star. First Slipher zeroed in on his target with the Lowell Observatory's twenty-four-inch refracting telescope. Then he passed the light through a prism that split the light into its separate colors and looked for characteristic markings within the spectrum: darker markings produced by, say, light absorbed by wisps of gas around stars or by the atmosphere around Venus. These markings identified the various elements in the objects, following the same technique that Kirchhoff and Bunsen had used a half century earlier to determine the composition of stars. Finally, Slipher compared the apparent color, or wavelength, of those markings with a reference spectrum of a stationary source in a laboratory. The Dopper shift, the amount of displacement in the spectrum, directly indicated how quickly the object was moving toward or away from the earth. Slipher spent much of his first decade at the Lowell Observatory picking apart the rays from Mars and other planets. He sniffed out the composition of their atmospheres and tried to measure the rotation of Venus and Uranus. Then he was ready to tackle the much fainter glow of the spiral nebulae.

Assuming the nebulae were spinning clouds of gas, Lowell expected Slipher would find one side of the spiral blueshifted, or approaching, and one side redshifted, or receding. Slipher discerned such turning motions all right—galaxies do, after all, rotate—but little else that dovetailed with Lowell's thinking. First of all, the spectra looked like those of stars, not of gas clouds. Second and more surprising, the nebulae as a whole were moving at breakneck speeds. In 1912, Slipher analyzed his first spectral measurement and discovered to his amazement that the Andromeda nebula is racing earthward at two hundred miles per second, much faster than any other object previously studied. Now Slipher started changing his focus, "investigating not only the spectra of the spirals but their velocities as well." Andromeda turned out to be an anomaly, which leads to the third and most unexpected discovery. An overwhelming majority of the nebulae he observed had their spectra shifted to the red. They were retreating from us at speeds of up to seven hundred miles a second, or some forty times the speed at which the Earth circles the sun.

Slipher's colleagues were bowled over by these giddy nebulae. When he finished reading his report at the 1914 meeting of the American Astronomical Society in Chicago, the audience gave him a standing ovation. Even among professional scientists, who are more easily brought to tears by spectral data than most ordinary people, this was a rare show of enthusiasm. At first the astronomers thought Slipher's measurements might reveal the net motion of the Milky Way with respect to the nebulae. That hope dimmed as Slipher pressed on with his study. By 1917, he had examined twenty-five galaxies and found twenty-one of them were redshifted; if we were drifting past them, half should be shifted to the blue. From then on out, every new galaxy showed the same trend. When Slipher wrapped his project in 1926, having tapped out the Lowell Observatory's modest telescopic power, his tally read four nebulae approaching, forty-one receding.

These runaway nebulae were moving so swiftly that they would quickly escape from our galaxy if they had started out nearby. It seemed only logical, then, that they had never been a part of the Milky Way in the first place. Slipher's 1914 presentation and his follow-up reports therefore buoyed the spirits of those who had never lost faith in Herschel and his belief in a universe full of other island universes. Hertzsprung praised Slipher for his work and took it as proof that the spiral nebulae are separate systems, each comparable to our own. Hertzsprung's support carried particular authority, as he had just completed his demonstration on the use of Cepheid variable stars for determining the distances to cosmic objects. Using his new calibrations, he had deduced that the Magellanic Clouds, hazy patches of light in the southern sky, are independent, satellite galaxies of the Milky Way. Slipher's findings on the nebulae dovetailed perfectly with that view.

Slipher stopped short of declaring that he had solved the riddle of the spiral nebulae. In private, he had no doubt that these nebulae must be other galaxies, but he lacked the absolutely crucial information to prove this interpretation: He did not know how far away they were. Because he couldn't say where the nebulae were, he also couldn't say for sure *what* they were. He couldn't rule out the possibility that they were small, nearby objects shooting out from our galaxy at high speeds. Einstein could grab the entire universe with his equations but Slipher, saddled with a smallish telescope, could only grope at the periphery of the glory of sci/religion.

Similarly, Slipher couldn't determine the meaning of the enormous and ubiquitous motions of the spiral nebulae. Only a few of the brightest ones, most notably the prominent Andromeda nebula, appeared to be approaching. The fainter ones all showed the strong redshifts indicative of receding objects, and the faintest spirals were moving fastest of all. From this pattern, Slipher was on the verge of discovering that the speeds of the nebulae are proportional to their distance, the indelible signature of an expanding universe. But again, the lack of any way to measure the distances to the nebulae stopped him in his tracks. Slipher had neither the equipment nor the kind of training that could give him the final bit of information, that clinching evidence that would have inscribed him in the history books. So despite the hubbub surrounding Slipher's work, some of the sharpest astronomers of the day still sided with Shapley and refused to accept the existence of other galaxies.

Looking back, the gap between theory and observation is painfully obvious both in the Great Debate and in Slipher's studies. For the moment, the visual astronomers didn't have the conceptual tools they needed to extend the reach of their telescopes. Einstein, on the other hand, had all the reach he needed—he had effortlessly spread the power of general relativity across one hundred million light-years of unknown space—but lacked the numbers that would give his cosmological ideas full sci/religious authority. Had he seriously questioned his colleagues, he would have very likely learned of Slipher's observations—they had, after all, been the hit at one of the world's leading astronomy conferences. He might then have paused before insisting on his static universe. Clearly there was more out there than the slow-moving stars of the Milky Way. But he was seeking God in the equation and was content to let others work out the details of his divine vision.

Amazingly, this revolution didn't take very long. Four years after the Great Debate, the question of "the scale of the universe" was decisively settled, Slipher was vindicated, and Shapley's arguments about the spiral nebulae were swept aside. Within a decade, Einstein's static universe was defeated once and for all, and science got into the creation business. All it took was the work of a latter-day Galileo, a scientific heavyweight named Edwin Hubble.

CHAPTER FIVE

EINSTEIN'S PROPHECY FULFILLED

IF EINSTEIN WAS the Jesus of the new sci/religion, Edwin Powell Hubble was its Martin Luther. While Einstein meditated on his equations, Hubble squinted rigorously at specks of light captured on his photographic plates. While Einstein pondered "the Old One's secrets," Hubble sternly argued, "Not until the empirical results are exhausted need we pass on to the dreamy realms of speculation." But in the end it was Hubble more than any other who realized Einstein's mystical goal of bringing all the stars and all of space into the domain of sci/religious understanding. Following his observational approach to cosmology, Hubble set out toward the boundaries of Einstein's finite universe. Then he went a step beyond, collecting the key evidence that the universe is also bounded in time—that astronomers might be able to reconstruct the origin of the world.

Hubble was a scientific juggernaut who flattened all in his path. During the Great Debate, Harlow Shapley had been able to claim persuasively that the Milky Way is the only galaxy in the universe. Within four years Hubble obliterated this view and reframed our galaxy as one of thousands, perhaps millions. Vesto Slipher had slaved for more than a decade to understand the motions of the spiral nebulae. Three years after he abandoned his research, Hubble cracked the code of the nebulae and collected the decisive evidence that our universe is expanding. After this bombshell, Einstein retreated from his static cosmology and eventually denounced Lambda as his "biggest blunder."

The man behind these heroic victories was a dashing enigma: a methodical observer, dogged and frequently brilliant, who was also a pompous celebrity hound, as inclined to magnify his accomplishments as to magnify the stars. Hubble's intricate, extremely private personality has confounded the most dogged biographers. Historian Gale Christianson, who has written an admirably thorough treatment of Hubble's life, lamented that "the astronomer revealed almost nothing of his inner universe." Again in a pointed contrast to Einstein, Hubble kept his motivations opaque. His acute instincts guided him to the most fruitful problems in astronomy. He worked his way to Mount Wilson, then the greatest observatory in the world, where he rapidly emerged as the star player among a brilliant team. Yet all the fame and professional success were not enough. He worried that others would try to take some credit for the discoveries he claimed as his own. And he was forever plagued with insecurities that some of the fabricated details of his past would come to light and humiliate him. As a result, he never wrote a memoir, and his wife, Grace, spent years after his early death in 1953 editing his letters and her journals to make them accord with his official history.

Hubble claimed to have rejected a lucrative job as a lawyer; in actuality, he never passed the bar. He was a competent heavyweight boxer in college, but his much repeated tale that promoters in Chicago urged him to go professional was most likely a youthful fantasy. The objectivity of Grace's recollections can be judged from her description of the first time she saw her future husband: "The astronomer looked an Olympian, tall, strong and beautiful, with the shoulders of the Hermes of Praxiteles . . . there was a sense of power, channeled and directed in an adventure that had nothing to do with personal ambition and its anxieties and lack of peace." Fortunately, she is not the only source of information about him. Edwin Hubble was born in 1889 in Marshfield, Missouri, the son of a lawyer. At the University of Illinois he studied astronomy under Forest Ray Moulton, a leading astronomer of the day, who was convinced the spiral nebulae were the birthplaces of stars. Hubble grew determined to solve the riddle of the nebulae but avoided mentioning this intention to his family, who counted on his finding a career in law. He continued on as a Rhodes scholar at Oxford University, where he majored in jurisprudence. Although his interest in the legal practice did not stick, the culture of Oxford did. The man from Missouri retained a penchant for British attire and British pronunciations—spitting out an occasional exclamation of "Bah Jove!"—for the rest of his life. Upon returning home, he taught Spanish and coached high school basketball in New Albany, Indiana, seemingly uninspired by the career choices that lay ahead.

With Moulton's encouragement, Hubble returned to the University of Chicago to earn a Ph.D. in astronomy, but he got sidetracked again. Just as he was offered a position at Mount Wilson, the United States entered World War I and Hubble enlisted. He asked George Ellery Hale, the observatory's director, to hold the job for him. Hubble knew that his studies of the nebulae would depend on access to a powerful telescope, and there was none more powerful than the one-hundred-inch reflector about to begin service on Mount Wilson. Perched atop the chalky mountains near Pasadena, 5,700 feet above sea level, Mount Wilson boasted crystal-clear air and a relaxed commute to the observatory's nerve center at nearby Caltech. The one-hundred-inch Hooker telescope—funded by businessman John Hooker and industrialist turned philanthropist Andrew Carnegie—was the premier example of America's burgeoning wealth and scientific prestige. Even its primary mirror had a noble pedigree: it was cast from green wine-bottle glass in the same factory that had fabricated the mirrors at Versailles for King Louis XIV. Much to Hubble's relief, Hale agreed to keep the position open. Hubble then set off the battle, having hastily completed a dissertation on the classification of faint nebulae by form, brightness, and size. Meanwhile, Shapley remained behind at Mount Wilson, analyzing the languid pulsing of Cepheid variable stars in order to trace the outlines of the Milky Way.

In pointed contrast to Einstein, Friedmann, and Lemaître, Hubble found the military life exhilarating. He was forever thrilled at being called "Major," perhaps because he had been spared the worst horrors of war. Although he served in France, he never saw any serious combat action, much to his naive regret. "I barely got under fire and altogether I am disappointed in the matter of the war," he wrote to Edwin Frost, his adviser at the University of Chicago's Yerkes Observatory. After the war he stayed on for a while in England, which afforded him the opportunity to sit in on lectures by Arthur Eddington, cosmology's great matchmaker, at Trinity College. Finally he could delay the inevitable no longer. In 1919 he returned to life in the United States and, more happily for him, to the job waiting for him at the new observatory. By Christmas he was puffing his pipe in the darkness, working the guide controls of the new Hooker telescope among the pine forests of Mount Wilson.

Hubble quickly established a reputation as punctual, patient, and vain—he still swaggered around in jodhpurs and military boots after starting work at the observatory. He clashed with Shapley, who, like Hubble, was born in Missouri (just seventy miles from Hubble's birthplace) but was worlds apart in his outlook. In the Great Debate, Shapley felt anxious at having to make a public presentation to an audience unfamiliar with astronomy. Here at Mount Wilson he was in his element. Combative, antiwar, and determinedly down-to-earth, Shapley was Hubble's bête noir. While Hubble served in Europe, Shapley had developed a reputation as one of Mount Wilson's top observers. His prestige, and his ego, presented natural obstacles to Hubble, though not for long. Immediately after the Debate, Shapley gambled and took the director's position at Harvard College Observatory. The job brought great prestige, but it meant losing access to the bucketfuls of starlight gathered up each night by Mount Wilson's one-hundred-inch mirror. If any telescope could expose the identity of the spiral nebulae and establish the true construction of the cosmos, surely this was it. According to one story, Shapley had seen photographs of the Andromeda nebula that revealed variable stars, proving the nebula was actually a distant galaxy, but he refused to believe the evidence before his eyes. Once he left there was no second chance. The job now fell to Hubble.

At Mount Wilson, Hubble returned to basics and began consolidating his earlier work on the classification of nebulae. Photographic surveys kept turning up more and more of these fuzzy blobs. By now the celestial catalog listings bulged with tens of thousands of nebula entries. There were "planetary" nebulae, named by William Herschel because their round shapes suggested ghostly echoes of the disks of the planets when viewed through the telescope. There were irregular smears of incandescence that intermingled with stars. There were dark patches that blotted out the light of the Milky Way. And then, of course, there were the infernal spiral nebulae that went their own way. To complicate things further, Slipher had found that some nebulae shine by reflected light, so their spectrum naturally looks much like that of a star. That discovery somewhat negated the commonsense argument that spiral nebulae had to be galaxies because their light appeared so much like that of a mass of stars. On the other hand, Slipher had found that the spirals appear to be receding at enormous velocities, which he considered near conclusive evidence that they are separate systems similar to the Milky Way. The Great Debate had produced no immediate movement on this issue. Eager but tentative,

Hubble classified the spiral nebulae separately from the others without overtly endorsing the still-controversial "island universe" theory.

Part of Hubble's genius lay in tackling the right problem at the right time, and his instinct told him that following the spiral nebulae would lead to a scientific bonanza. But he was operating on more than pure intuition. In 1922—the same year Alexander Friedmann developed his first mathematical description of an expanding universe—the Swedish astronomer Knut Lundmark observed what he believed were individual stars in the arms of M33, a bright spiral nebula in a small, faint constellation with the dryly geometric name of Triangulum. Shortly thereafter, John Duncan at Mount Wilson spotted dots of light that grew fainter and brighter in the same nebula. To all appearances these were variable stars, similar to ones in the Milky Way but far dimmer owing to their great distance.

Sensing the answer was at hand, Hubble stepped up his efforts. He spent long nights on his bentwood chair, guiding the movements of the riveted-steel mount of the Hooker telescope to cancel out the Earth's rotation and stay true to the stars. The effort paid off with highly detailed, long-exposure images of the Andromeda nebula. Now the mottled light of Andromeda began to resolve itself into a multitude of luminous points, not a smear of gas but a vast hive of stars. Clinching proof came in October of 1923, when Hubble found the telltale flickering of a lone Cepheid variable among the grainy stellar multitude in one of Andromeda's arms. He watched the star's brightness peak, then dip, then rise again on a thirty-one-day cycle. Hubble then looked up the relationship between period and luminosity for Cepheids, which Shapley had refined while mapping the Milky Way, to derive the distance to Andromeda. His estimate was 930,000 light-years, less than half the current value, but a shockingly large number at the time. That distance placed Andromeda, one of the brightest and presumably the nearest of the nebulae, far outside Shapley's "big galaxy" model.

The Great Debate was settled then and there. Spiral nebulae were other galaxies, and the universe was every bit as vast as Einstein had imagined it to be. Hubble had gambled that the universe would submit to scientific inquiry, and he had won. Still he pressed on. By the following February, he had uncovered nine novas and a possible second Cepheid in Andromeda. He also found Cepheid variables in Triangulum and possibly in three other nebulae as well. Now that there could be absolutely no doubt of the meaning of his finding, he wrote to Shapley to needle him with the news. "You will be interested to hear that I have found a Cepheid variable in the Andromeda Nebula," the letter began. Shapley needed to read no farther to understand the significance of Hubble's words. In his reply, Shapley raised a number of skeptical and cautionary arguments, but privately he despaired. "Here is the letter that destroyed my universe," he morosely told Cecilia Payne-Gaposchkin, now a doctoral candidate at Harvard, who was in his office when Hubble's missive arrived. By an odd coincidence, Payne-Gaposchkin had also been present when Eddington announced the results of the eclipse results that confirmed general relativity, making her witness to two pivotal cosmological advances in the space of four years—a sort of Mary Magdalene in the shadows of the sci/religious miracles.

Once again, the size of the measurable universe had tremendously increased. Until the middle of the nineteenth century, the most remote identified object was Uranus, roughly two billion miles away. When Friedrich Bessel nailed down the distance to a star for the first time, he placed 61 Cygni ten light-years away, thirty thousand times farther than Uranus. Now Hubble had determined the first galactic distance and moved the boundary of measured space another one hundred thousand times farther out into the depths of space. The scale of human life shrank each time, while the evident power of the human intellect grew. Hubble had penetrated the secret depths of space where Shapley had imagined that true galaxies hid, like blue whales lumbering at uncharted fathoms in the murk of the oceans. If the God of the old-time religions was out there, he had fewer and fewer places to hide. But if knowing the universe is the same as knowing God, as Einstein preached, then astronomers were closing in on the divine.

Despite his obvious excitement at the Andromeda findings, Hubble was reluctant to publish his results. For all his surface confidence, he was terribly concerned about making a grand pronouncement that would appear naive or foolish. Every time he walked down from the summit to attend the formal 5 P.M. dinners at the Monastery, Mount Wilson's living quarters, Hubble had to face his astronomer brethren. Not all of them accepted the existence of other galaxies. Adriaan van Maanen, a playful and well-liked Dutch astronomer, had in fact argued vigorously in the other direction. He was convinced that he had observed some of the spiral rotating, which was possible only if they were relatively small and nearby. Indeed, one of the reasons Shapley refused to believe the spiral nebulae could be galaxies was that he completely trusted these rotation measurements. Unlike Shapley, Hubble didn't consider van Maanen a good friend. Nevertheless, he found it worrisome to have a doubter in his own midst and held back until he was utterly sure of his results. (Van Maanen never figured out where he went wrong and refused to admit his errors long after Hubble had proven conclusively that spiral nebulae are galaxies. In a fit of pique, Hubble reexamined his photographic plates and declared that "the large rotations previously found arose from obscure systematic errors and did not indicate motion, either real or apparent, in the nebulae themselves.")

Word of Hubble's discovery inevitably leaked out to the media. As a result, the first public announcement of his astronomical breakthrough was a small story that ran in *The New York Times* on

November 23, 1924. Still Hubble balked at publication. The noted stellar astronomer Henry Norris Russell pressed him to present his findings to a Washington, D.C., meeting of the American Association for the Advancement of Science, which offered a $1,000 prize for best paper. When he didn't submit anything, Russell snorted, "Well, he is an ass. With a perfectly good thousand dollars available, he refuses to take it," then turned to find that Hubble's paper had just arrived. Hubble remained in splendid isolation at Mount Wilson while Russell read his official announcement to an enthusiastic crowd on January 1, 1925. Hubble shared the best-paper prize.

Oddly enough, it was Shapley, not Hubble, who suggested that astronomers should adapt their nomenclature to the new reality and call the external star systems "galaxies." Like many revolutionaries in the new church of sci/religion, Hubble still carried within him the conservative views of the world he overthrew. He was also naturally inclined to disagree with any idea that came from his rival, Shapley. Hubble, the man who proved that the Milky Way is but one of innumerable galaxies, forever called the objects by the archaic name "extra-galactic nebulae."

As Hubble watched the cyclical flaring and dimming of the Cepheids in Andromeda, he did more than establish the distance scale of the universe. He also erased the lingering concern, aired by Curtis in the Great Debate, that stars lying at great distances from us might behave differently from those in our immediate celestial neighborhood. And now that scientists could determine the distances to other galaxies, they could establish the constancy of the universe over time as well. If the Andromeda galaxy is one million light-years away, that means the light we see now started on its way earthward a million years ago. That is, we are seeing the stars in Andromeda as they were a million years ago—yet they look identical to nearby stars. As Hubble and other astronomers looked out to ever greater distances, they added ever more evidence for the principle of temporal uniformity. This constancy of nature lent credibility to the search for a single set of overarching cosmic rules. Or, as Einstein might have put it, it showed that God does not change the house rules of the cosmos.

What Hubble did not settle was the lingering question of whether Einstein was correct when he assumed a static universe, held in place by Lambda. De Sitter had shown that Einstein's cosmology was not the only possible interpretation of general relativity. Friedmann had issued a challenge in suggesting that Einstein's own equations implied a dynamic universe. And by 1925, Slipher had collected spectral data on forty-one spiral nebulae, almost all of which showed a strong redshift. This perceived reddening of light implied that the sources were moving away from us at hundreds of miles per second; the record holder was racing 1,100 miles farther away every second. As long as the nebulae could be dismissed as bits of flotsam breaking free from the Milky Way, their breakneck speeds didn't seem such a vital matter. Now that Hubble had demonstrated that each smudge of light was another galaxy as mighty as our own, the redshifts took on a much greater significance.

If there was a clear pattern to the galactic redshifts, that would spell trouble for Einstein's cosmology. His static universe did not alter light from distant bodies and did not allow for any large-scale motions in the universe. The de Sitter universe, aka "solution B," was in better shape—it predicted that light might grow redder as it passes through space, in which case galaxy redshifts should increase in proportion to their distances. But de Sitter didn't interpret his redshifts as velocities, and even he didn't consider his mass-free universe anything more than a mathematical idealization of the real world. That left a third possibility, that the redshifts really represented a true, systematic motion of galaxies away from us. The notion of a literally expanding universe was too weird for Friedmann, whose work was at the time still unknown to most of the astronomical world anyway. All the same, what creation had written in the book of nature seemed to be suggesting just that, even before Hubble got to work.

As early as 1921, astronomer Carl Wirtz in Germany thought he saw signs of a proportional relationship between distance and redshift in Slipher's observations. Ludwik Silberstein, a Polish-born physicist then working in England, stirred up the pot in 1924, when he alleged to have proof of such a link. In making his claim, however, he selectively included studies of star clusters in our own galaxies. His colleagues, smelling a rat, quickly denounced the results, and many of them soured entirely on the provocative but endlessly ambiguous redshift studies. Undaunted, de Sitter continued to search for evidence of his cosmological reddening effect. Early in 1928, de Sitter had a chance to discuss the matter with Hubble, who was traveling through Europe. Hubble had read about de Sitter's research and was persuaded by the distinguished theorist's entreaties to unleash Mount Wilson's huge Hooker telescope and resolve the mystery of the redshifts.

Again, Hubble's timing was perfect. Slipher, with his limited resources, had run into impenetrable theoretical and practical barriers by 1926. Analyzing the light of spiral nebulae was slow business, especially using the Lowell Observatory's modest twenty-four-inch refracting telescope. He would expose the same photographic plate over several nights, gradually building up a total exposure time of twenty hours or more. It was tedious work, and when he was done he could still only guess whether faraway objects were moving more quickly than nearby ones, because he had no way of measuring the distances to these nebulae. Hubble or Shapley could look at Cepheid variables because they had access to a one-hundred-inch mirror; such work was way out of the Lowell Observatory's league. Poor Slipher was reduced to looking at the sizes and brightnesses of the nebulae in order to guess their relative proximity. In his heart he believed these objects must be other galaxies. But he had pushed his equipment to the limit, and still he could not obtain the final answer.

Hubble confidently picked up where Slipher left off. He approached the challenge armed with two potent weapons: Mount Wilson's mighty one-hundred-inch telescope, which collected seventeen times as much light as Slipher's instrument; and Milton Humason, the observatory's crackerjack photographer. In his teens, Humason had worked at Mount Wilson as a mule driver, helping to deliver construction materials for the Monastery and other buildings around the observatory. Bewitched by the temple of astronomy taking shape high up in the San Gabriel Mountains, he returned there in 1917 as a janitor but proved himself curious and capable of much greater things. He started helping out one of the students, then began assisting Shapley, and finally he graduated into one of the finest photographers at the observatory. Hubble picked out the images and spectra he needed, and time and again Humason delivered, guiding the telescope and developing the photographic plates to perfection.

To get any kind of meaningful perspective on the spiral nebulae, now properly known as spiral galaxies, Hubble knew he had to obtain accurate distances and redshifts for a large sample of them. That meant going deep, well beyond the handful of nearby bright spirals. Starting in 1928, Hubble and Humason trained Mount Wilson's unblinking one-hundred-inch eye on two dozen of Slipher's galaxies, confirming and extending the dominance of redshifts that Slipher had recorded. Then Hubble did what Slipher could not do—he calculated the distances to those galaxies. For the nearest ones, he monitored Cepheid stars like the ones he saw in the Andromeda galaxy. Probing deeper into space, he looked for nova explosions or the most brilliant giant stars in each galaxy, which are all roughly equal in luminosity. Farthest out, Hubble noted that the brightest galaxies in large clusters all seemed quite similar, so he used them as crude "standard candles" for gauging distance. This clever stepping-stone approach allowed him to locate galaxies a hundred times more distant than Andromeda. By 1929 he had collected redshifts for forty-six galaxies and could tell the distances, more or less, to two dozen of them.

Finally, Hubble made a graph showing how the velocities—as indicated by how strongly the light is shifted toward the red end of the spectrum—relate to the distances of the galaxies. The graph showed a straight line: the farther away the galaxy, the more quickly it recedes. This linear relationship, now called Hubble's law, is the signature of an expanding universe. Princeton cosmologist James Gunn summed up this work: "Hubble's gigantic realm had been endowed with motion, motion which implied physical process, evolution, and origin." Einstein delivered the prophecy of a dynamic, all-encompassing cosmic theory. Now Hubble witnessed the miracle that had been foretold and testified about it before the world.

Because the galaxies seem to move outward from us in all directions, it might seem as if we are in some unique, central location. But in an expanding universe, that is what every observer sees; it is precisely for that reason, in fact, that scientists quickly interpreted Hubble's law as proof of such cosmic expansion. Consider again the analogy in which Einstein's curved space is represented by the surface of a rubber balloon. Einstein had assumed that the balloon remained perfectly still. But suppose instead the balloon was expanding so that over a period of time its size doubled. What would an observer—our dear old hypothetical, two-dimensional friend Trevor—see from his tiny perch on the balloon's surface? A spot one inch away would, after the period of doubling, be two inches away. A spot two inches away would end up four inches away. A spot three inches away would end up six inches away, and so on. In other words, Trevor would observe each point on the surface of the balloon moving away at a rate exactly proportional to its distance. That is what happens when every part of the balloon (representing every part of space in the real world) expands at an equal rate. If the real universe were expanding, the same kind of thing would happen. Light from distant galaxies would be stretched and reddened, and the intensity of the effect would be in direct relation to how far away each galaxy is.

Hubble, ever the cautious researcher, dared not come right out and say that the universe is expanding. He merely set out his findings in a paper, soberly titled "A Relation Between Distance and Radial Velocity Among Extra-Galactic Nebulae," that appeared in the March 15, 1929, issue of *Proceedings of the National Academy of Sciences.* It was just six pages long, terse yet confident. His actual data points were scattered all over the page, more like potshots than a scientific bull's-eye. There were large random errors in his distance measurements, as well as large random galactic motions that distorted the pattern. He drew a hopeful line through the data points, depicting the linear relationship between distance and redshift that he sensed was there. "For such scanty material, so poorly distributed, the results are fairly definite," he wrote unapologetically. Through a mix of inspired genius and dumb good luck, he saw the correct pattern in the clutter.

Given what it spawned, Hubble's paper was surprisingly stingy on big concepts. "This discovery finally brought the question of the beginning of the universe into the realm of science," writes Stephen Hawking. But Hubble didn't talk about beginnings. He didn't talk about the expanding universe. He didn't even talk about galaxies and motion. The physical implications of his redshifts lay outside his empirical conception of science. Eight years later, after most of his colleagues had firmly converted to sci/religion and rewired their brains to accept the idea of a cosmos that grows, Hubble still balked. "Well, perhaps the nebulae are all receding in this peculiar manner. But the notion is rather startling," he said. Hubble was not a theorist. He understood little about general relativity, and he knew nothing of Friedmann's dynamic cosmological models. Here again, Hubble comes across as the near opposite of Einstein, although Hubble's Puritan adherence to observation echoes the philosophy of Einstein's onetime hero Ernst Mach. Mach was so allergic to speculation that in 1906 he was still writing about "the artificial hypothetical atoms and molecules of physics and chemistry." Although less extreme in his views, Hubble was generally content to attempt to comprehend the universe through data alone. He stuck to his austere Lutheran approach, reading nature's holy scripture without constructing his own arguments.

Nevertheless, Hubble sensed that his role as the grand explainer of the universe required him to comment on theoretical interpretations, no matter how alien they seemed. Hubble made one awkward attempt at the end of his 1929 paper to connect his observations with what little he knew of cosmological theory. "The outstanding feature, however, is the possibility that the velocity-distance relation may represent the de Sitter effect, and hence that numerical data may be introduced into discussions of the general curvature of space," he wrote. Quite likely he viewed this endorsement as a kind of repayment to de Sitter for having shared his theoretical ideas and encouraging Hubble to pursue the redshift measurements. The de Sitter universe also appealed to the conservative in Hubble, because it explained the redshifts yet in its abstruse way maintained a classical calm. Too bad for Hubble, he was rather late to this party. Nobody else was sold on this confusing and unrealistic formulation, and de Sitter abandoned it not long after Hubble presented his findings.

Such flubs only deepened Hubble's aversion to revealing the theoretical or philosophical goals underlying his research. Many current observers still follow in this tradition. They aspire to a kind of mastery of the universe, itching to know the most distant, obscure details of the construction of the cosmos. Yet if pressed, they swear they don't have a mystical bone in their bodies. Hints of Hubble's yearnings break through his wall of self-censorship. Near the end of *The Realm of the Nebulae,* his book summarizing his telescopic explorations, he addressed the limits of his explorations: "With increasing distance, our knowledge fades, and fades rapidly. Eventually we reach the dim boundary—the utmost limits of our telescopes. There, we measure shadows, and we search among ghostly errors of measurement for landmarks that are scarcely more substantial." Here Hubble expressed his dark spiritual fear, that his telescopes would never be able to penetrate the kingdom of God. He never went to church and professed not to have any personal faith, but Hubble's worries were those of a believer. Empirical observation was his Lord, and the limits of his telescopes represented the grim end points in his search for truth.

But in 1929 Hubble was nowhere near those limits. By the time the first velocity-distance paper hit the press, Humason was already busy collecting more spectra at machine-gun pace. The new observations amply confirmed the linear trend of Hubble's law. In 1931, Hubble and Humason published a follow-up paper that added fifty new galaxies to the graph of redshift versus distance, including a galaxy cluster in the constellation Leo they estimated to be more than hundred million light-years from the Earth, moving away at twelve thousand miles per second. The signs were all around: the universe is expanding ecstatically, and Einstein was wrong to think he needed Lambda to hold it in place. (An inquisitive reader might well ask why the galaxy, the solar system, the living room, or this book is not caught up in this overarching expansion. The answer again lies in Einstein's equations. Gravity counteracts the stretching of space, so all these objects remain intact pretty much indefinitely. It is only in the regions between galaxies, where matter is scarce and space itself dominates, that the expansion occurs.)

Although Hubble's empirical results spelled doom for Einstein's Lambda-dominated, immobile universe, they strongly endorsed another key aspect of his theory, the cosmological principle.

Where his earlier work affirmed the uniformity of physical law throughout the universe, Hubble's later work demonstrated the uniformity of matter that Einstein's theory demanded. The farther Hubble looked, the more galaxies he saw. Everywhere the pattern was the same. Galaxies might congregate in small groups or large herds, but on the largest scale they were spread evenly through space. As his successors have surveyed greater depths of space, this distribution continues to hold. Likewise, the cosmic expansion appears pervasive; no corner is exempt. Every spot in the universe seems much like every other.

This uniformity, which Einstein called "the cosmological principle," has become an enshrined canon of astrophysics. It means that we do not live in a privileged location in the universe. What we see is, overall, the same as what any observer in any other spot would see. Conversely, we can extrapolate from our observations of nearby regions and assume things are generally like that everywhere else as well. In essence, this is a modern elaboration of Descartes's faith that he could trust the evidence of his senses because God would not set out to deceive him. Without the cosmological principle, Einstein could not securely apply the equations of general relativity to the universe as a whole. So while Hubble was blasting apart one detail of Einsteinian cosmology, he was vindicating its fundamental underpinnings. Although the cosmological principle greatly simplified the task of making a mathematical model of the universe, it caused headaches for Einstein's successors as they tried to reach even deeper and ask *why* the cosmos should be so uniform.

But Einstein was neither celebrating nor cursing. During the 1920s, while Hubble was busy redrawing the universe, Einstein's attention was elsewhere. He searched madly for a grander version of general relativity that would unify gravity with the seemingly unrelated laws governing electromagnetism, his "unified field theory." At the same time, he grumpily disputed the growing number of scientists who believed quantum physics proved that the world operates according to the rules of chance, not by absolute cause and effect. He also started to involve himself with the burgeoning Zionist movement.

The task of integrating Einstein's general relativity with Hubble's zooming galaxies fell to Georges Lemaître, the Belgian abbé who moved effortlessly between the society of clerics and that of cosmologists. He expressed faith in the simplicity and beauty of the scientific world as passionately as he pursued the possibility of finding salvation in the religious world. By the time he published his first cosmology paper in 1925, Lemaître had already moved beyond de Sitter's idealized cosmology and started to realize that the natural state of a universe ruled by general relativity is to shrink or to grow. He thought of "solution B" not as an empty void, but as the end point of a universe that has expanded so much that its matter is diluted almost to nothingness. From his visits with Slipher and others, he knew that the spiral nebulae appear to be fleeing from us at enormous speeds. He sought to explain this motion in terms of real physical change, which is why he spoke of the "non-statical character of de Sitter's world."

In 1927, Lemaître published a paper summarizing his refined cosmological ideas. He set out to build a universe "intermediate between that of Einstein and de Sitter." Like Einstein's, it contained matter; like de Sitter's, it explained the reddening of the nebulae. But unlike either, the Lemaître universe was devoted to modeling physical details rather than tending to the philosophical ideals of Lambda or the mathematical abstractions of the de Sitter effect. Lemaître spoke of galaxies, not theories of inertia or test particles. Drawing on his extensive familiarity with astronomy and thermodynamics, Lemaître considered the effects of radiation pressure and temperature changes and regarded cosmic expansion as a true, observable consequence of the way the universe arose and evolved. "The receding velocities of the extra-galactic nebulae are a cosmical effect of the expansion of the universe," he asserted. In his paper, he even speculated about the first cause of the expansion. Perhaps, he wrote, "the expansion has been set up by the radiation itself"—a bright, Genesis-like flash of a beginning, the first glimmer of what would later evolve into the big bang theory.

Almost as remarkable, Lemaître estimated the rate of expansion of the universe in his 1927 paper, two years before Hubble published his first results. How Lemaître did so is unclear. Evidently he performed his own analysis of published and unpublished data on the distances to various galaxies whose redshifts had been measured by Slipher and others. The number that Lemaître came up with (approximately one hundred miles per second of velocity for every million light-years of distance) was in fact very close to the value Hubble published two years later. Helge Kragh, a Norwegian historian of science who has championed Lemaître's work, considers this good evidence that Lemaître deserves credit for uncovering the distance-velocity relationship. "The famous Hubble law is clearly in Lemaître's paper. It could as well have been named Lemaître's law," he argues.

Unfortunately, both Lemaître's and Hubble's early calculations of the cosmic expansion rate contained considerable error. Because of a lingering misconception in how to interpret the pulsations of Cepheid variable stars, Hubble severely underestimated the distances to galaxies. As a result, his assessment of the expansion rate—which is just velocity divided by distance—was way too high. Taken at face value, the numbers implied a universe hardly more than a billion years old, which was absurd. From studies of the decay of radioactive elements, scientists knew the Earth was at least two billion years old. Some theoretical interpretations of the dynamics of stellar clusters implied our galaxy was much more ancient still, several trillion years old. Lemaître wasn't concerned, however, because he considered the present expansion merely a transient state of affairs. In his view, the universe had started out compact and static, resembling the Einstein model. At some point the whole became unstable, perhaps because of radiation pressure, and began to expand. The universe would then keep growing without limit until it thinned out into something resembling the de Sitter model. Thus Lemaître managed to avoid completely contradicting his illustrious predecessors. He also nicely sidestepped, for the moment, the complicated mystical question of when the universe began.

Lemaître had formulated a creative, mathematically persuasive argument in favor of an expanding universe. But he ran smack into the same barrier that Friedmann had hit just a few years earlier: he had a devil of a time getting the world to notice what he had written. It didn't help that Lemaître published his paper in the *Annals of the Brussels Scientific Society*, not exactly a must-read in the pews of sci/religion. Seeking to publicize his work, Lemaître sent a copy of his paper to Arthur Eddington, but to no avail—Eddington later confessed he had either ignored the mailing or never noticed it at all. Lemaître's cosmic solution was still largely unknown in October of 1927, when he attended the Solvay Conference in Physics in Brussels and sought an audience for his theory. When he approached Einstein at the conference, he received little encouragement. "Your calculations are correct, but your physics is abominable," Einstein responded, still averse to any cosmological model that changed over time. So things stood until 1929, when Hubble let the cat out of the bag. Einstein's disciples quickly discovered that the expanding universe was not the abomination he had believed. Quite the opposite: it realized his beautiful prophecy of a unified cosmic theory.

The exegesis of Hubble's discovery began at a January 1930 meeting of the Royal Astronomical Society. There, Eddington conferred with de Sitter to ponder the theoretical implications of the swiftly moving galaxies. Neither of the well-known cosmological models seemed compatible with the new evidence. In his book

The Expanding Universe, Eddington describes the almost comically uncertain state of cosmology at the time. "Shall we put a little motion into Einstein's world of inert matter, or shall we put a little matter into de Sitter's Premium Mobile?" he wondered. Eddington, a devout Quaker with a spiritual streak a mile wide, sensed it was time to introduce some fresh thinking. He assigned a research assistant to scour the literature looking for any clever insights into the physical principles of an expanding universe. Lemaître soon heard of Eddington's search for enlightenment and fired off a letter calling attention to his now forgotten paper. This time Eddington read it and, suitably impressed, hastened to draw attention to the Belgian priest's ideas. He quickly wrote up a semipopular summary of Lemaître's cosmological ideas for the British journal *Nature* and shepherded a translated version of the 1927 paper into the *Monthly Notices of the Royal Astronomical Society* in 1931. Even de Sitter hailed Lemaître's "brilliant discovery, the 'expanding universe.'"

Lemaître's model was still far from a straightforward picture of an outward rush of galaxies. In the latest formulation, cosmic expansion still emerged slowly from an earlier equilibrium state, which could have existed almost forever depending on the value of Lambda one plugged into the equations. Lambda also controlled how much time had passed since the beginning of the disequilibrium that sent the universe inflating into its present state. But what really mattered was that Lemaître had stated, in scientific but quite unequivocal terms, that the universe in its present form originated at a particular moment. Einstein had declared that science could build a theory that would cover all of space, not just the corner of the universe that we can see. Now Lemaître was claiming all of time as well.

No scientist before him had the temerity to propose a scientific model that would reach all the way back to the origin of the universe. The idea seemed too spooky; it was a task for the people who kept track of who begat whom in the Bible. Eddington and many of the other brilliant minds grappling with the meaning of Hubble's runaway galaxies still recoiled from the obvious logical leap: if everything is moving apart now, it must have all been much closer together in the past, and at one point, it must have been all crowded together at a single point. Lemaître made it clear that he meant his theory literally. Eddington would have none of it. Writing in the journal *Nature*, he stated, "Philosophically the notion of a beginning to the present order of nature is repugnant to me."

Lemaître took these words as a challenge. Just weeks after Eddington's paper appeared, he started to formulate a complete picture of how the universe emerged from an initial state that he called the "primeval atom." Such a universe would have a definite age: "A general conclusion of the theory of the expanding universe is that the time-scale of evolution is much shorter than was thought previously," he wrote. Depending on how one adjusted the parameters—in other words, how one read the mind of God—the time of the "rupture of equilibrium" could be as far back as hundred billion years ago. In developing this model, Lemaître countered Eddington by distinguishing the primeval atom from the universe it evolved into. The rupturing of the primeval atom gave rise to galaxies and to a spray of radiation. The moment when this formative event occurred was "a day without a yesterday," as Lemaître put it.

Given Lemaître's other life as a practicing Catholic priest, many scientists and historians have naturally assumed his cosmology was intended as a modern retelling of the Book of Genesis. The evidence points toward a distinctly different, more complicated interpretation. True, he did attempt to defend Catholicism from a direct, atheistic attack. "There is no reason to abandon the Bible because we now believe that it took perhaps ten thousand million years to create what we think is the universe," he told *The New York Times* in 1933. "There is no conflict." But rather than expanding the authority of old-time religion, Lamaître was constricting it and, like Spinoza, rejecting the biblical conception of a willful God. He derided those who attempt to bring classical theology into their research: "Hundreds of professional and amateur scientists actually believe the Bible pretends to teach science. This is a good deal like assuming that there must be authentic religious dogma in the binomial theorem." Years later he took pains to explain that the primeval atom "leaves the materialist free to deny any transcendental Being. . . . For the believer, it removes any attempt to familiarity with God."

Lemaître conceived of the primordial universe as an atom in part because he imagined that such an object would operate according to quantum rules, in which the physical state of a system can never be determined with perfect precision. "I would rather be inclined to think the present state of quantum theory suggests a beginning of the world very different from the present order of nature. . . . If the world has begun with a single quantum, the notions of space and time would altogether fail to have any meaning at the beginning," he wrote. Thus, the state of the modern world could not be preordained from the start. Such a restriction leaves a place for God, but only a hidden God who depends on quantum physics to maintain his veil. Lemaître may have believed in salvation, but practically speaking, his faith didn't look much different from Einstein's scientific religion, which Einstein described as "a rapturous amazement at the harmony of natural law, which reveals an intelligence of such superiority that, compared with it, all the systematic thinking and acting of human beings is an utterly insignificant reflection." With the primeval atom hypothesis, Lemaître followed Einstein in the search for harmony between quantum physics and general relativity. His effort represented the spirit of Galileo more than the spirit of the church.

All the same, Eddington objected to Lemaître's hypothesis, which he found it "too unaesthetically abrupt." He balked at running the clock all the way backward to an explosive beginning and preferred his own "placid theory," based on the earlier formulation in which a semistatic universe gently slips into a state of expansion after some indefinite stretch of time. But Lemaître was committed to his explosive primeval atom and continued to develop the idea. He pictured this atom as a giant radioactive nucleus whose decay set all the present events in motion. Some fragments of the original fireworks would survive. Lemaître thought these might explain the existence of cosmic rays—energetic particles that shower down onto the Earth from space—which had been discovered in 1925 and were still poorly understood. At a time when the inner workings of the atom were still largely an enigma, and the atom bomb not even yet a wild conjecture, such speculations were not unreasonable. During the 1940s and 1950s, cosmologists abandoned this particular picture but found that nuclear physics is indeed indispensable for understanding conditions in the early universe; today cosmology and particle physics are inseparable partners. Before they got diverted into military matters, famed atomic researchers such as Enrico Fermi and Edward Teller lent a hand in imagining what the first moments of creation might have been like. The broad outlines of Lemaître's primeval atom hypothesis lodged in the minds of his colleagues and became, with much modification, the modern big bang theory.

Lemaître's primeval atom universe consolidated the sci/religion revolution Einstein had started in 1917. Until Hubble announced a progressive pattern of velocities among distant galaxies, many astronomers (and, even more, scientists in general) regarded cosmology as little more than a highfalutin brand of mathematical philosophy. Now cosmology was an essential tool for explaining an otherwise baffling observation. There were a few dissenters, most notably the irascible Fritz Zwicky at Mount Wilson. But overall, the astronomical community was galvanized by the new discoveries and rapidly converted to Einstein's faith in a unified theory of the cosmos, something that had seemed so arcane and remote just a decade earlier. Lemaître didn't say so directly, but his model also pointed to another solution to Olbers's paradox. In the Einstein model, the night sky is dark because the universe is finite in size. In the Lemaître model, the key point is that the universe is finite in age. If ten billion years has elapsed since the atom burst and galaxies began to form, then by definition we can see only galaxies that are less than ten billion light-years away. Even if there is more universe out there, it is irrelevant to us. Its light simply will not have had time to reach us.

While Lemaître remained true to the spirit of Einstein's 1917 manifesto, he inverted many of its details. In addition to being dynamic, it was finite in age but potentially expanded to infinite dimensions. It retained Lambda but used it as a destabilizing rather than stabilizing force. Yet during those heady years from 1927 to 1931, Einstein did not speak up to defend his cosmology. He let his disciples carry out the search for God in cosmic dimensions, while he set off in pursuit of divine truth at the other end of the scale. Now his primary goal was to find a way to bring together gravity, which was described by general relativity, and electromagnetism, which seemed to follow the completely independent rules of quantum physics.

This task was energized by his deep dissastisfaction with the direction of quantum theory. According to the uncertainty principle, articulated by German physicist Werner Heisenberg, there is an inherent fuzziness to the workings of nature. One time a decaying atom might spit out a particle in one direction, next time in another, seemingly at random. Even the particle's location is vague, described in probabilities rather than absolutes. Numerous experiments seemed to show that the uncertainty principle is correct, but Einstein was not impressed. In fact, he was infuriated. He was guided to relativity by his belief in ironclad determinism—and now some of the leading scientists of the day were claiming the most basic behavior of subatomic particles appears governed by statistics. "[Quantum] theory yields much, but it hardly brings us closer to the Old One's secrets. I, in any case, am convinced that He does not play dice," Einstein wrote in a 1926 letter to Max Born, his frequent confidant. This quote later morphed into the more concise "God does not play dice with the universe."

Einstein's search for a unified field theory also drew sustenance from his conviction that general relativity must apply everywhere in the universe. When he looked outward, this led him to his "Cosmological Considerations" and the first all-encompassing mathematical description of the universe. Whatever details he got wrong, he could rest assured that the basic approach was sound: relativity really does apply even at the largest scales. When he looked inward, however, there seemed no room for gravity. Three forces—electromagentism and the two nuclear forces, known only as "strong" and "weak"—dominate the atomic world. But gravity should operate there, too. If electrons behaved like planets circling the atomic nucleus, general relativity predicted that they should constantly shed tiny bits of gravitational energy. Gradually the whole system should run down. The only way to avoid this would be if gravity follows quantum rules, which forbid such energy leakages from happening. But Einstein had no quantum theory of gravity. Again, the two systems were at loggerheads.

Judging from the amount of effort Einstein expended and the meager progress he achieved, his quest to reconcile relativity and quantum theory was, if not his true greatest blunder, at least his greatest blind alley. In January of 1929, as Hubble was putting the final touches on his study of galaxy redshifts, Einstein published the first paper intended to show the world the unity in physics that, for the moment, he alone could see. Within a year, Einstein rejected his ideas as unworkable, moved from Vienna to Berlin, and began a collaboration with a young American physicist named Walther Mayer, whom he hoped would aid him in his holy quest.

Many more relocations, collaborators, and false starts lay ahead. In the early 1930s Eddington joined in, following his own idiosyncratic path. Decades later, the unified field theory remains a painfully unattained goal in physics.

Sweeping changes in the scientific and political realms shifted Einstein's interest back to cosmology, for a brief while, at least. In economically depressed Germany, Hitler's support was on the rise, and Einstein had become a prominent focal point for anti-Semitic attitudes. The publication of *100 Authors Against Einstein,* a torrent of cheap attacks on Einstein's ideas and character, was indicative of the changing political climate. Einstein spent much of 1930 abroad. When he returned to Berlin, he received a visit from Arthur Fleming, chairman of the board of trustees of Caltech, who invited him to take a temporary position as a research associate at the institute. Einstein, intrigued by reports of the momentous discoveries coming from Mount Wilson, which is operated by Caltech, required little persuasion. In December 1930, the Old World's champion theorist set off for a showdown with the upstart observers in the New World.

Einstein had scarcely kept up with theoretical physics over the past dozen years while he buried himself in his unified field theory. He was even less in touch with the state of deep-space astronomy. He was therefore eager to visit the California peak where, he noted, "new observations by Hubble and Humason concerning the red shift of light in distant nebulas make it appear likely that the general structure of the universe is not static." At Mount Wilson, the always status-conscious Hubble latched on to Einstein like a bulldog. He proudly walked his famous visitor through the mechanical workings of the huge observatory, showing off the spectrographs that had detected those wondrous redshifts. Einstein insisted on the full tour, including close-up examination of the telescope mechanism, spectrographs, and photographic plates. Throughout, the camera bulbs flashed and Hubble made sure to appear within the same frame as Einstein whenever possible. In a particularly proud moment, the Hubbles had the Einsteins over for dinner. Not taking any chances, Hubble had invited a young blond actress named Doris Kenyan, gambling that Einstein would be charmed by the presence of a genuine Hollywood starlet.

He was, but Humason's photographic plates carried out the greater seduction, convincing Einstein that the redshifts were real. Cosmology had changed drastically since Einstein dismissed Lemaître's ideas at the Solvay Conference back in 1927. Now he scrambled to catch up. On February 4, 1931, Einstein told the assembled media at the observatory that he officially rescinded his original cosmology and endorsed the expanding universe. Chastened by Humason's rapidly growing stack of galactic redshifts, Einstein renounced his static conception of the universe and pointed out that cosmic expansion fits in perfectly well with general relativity—something that had been true all along, of course. Then he glanced at his watch—he was running late, as usual—flashed another of his trademark indeterminate smiles, and darted out of the room, brushing aside the questions erupting from the dazzled swarm of reporters.

To the sci/religious faithful, this was not exactly fresh news. Hubble's measurements of galactic motions were well-known. Most of the scientists who were seriously following these developments had already recognized that the recent findings discredited the kind of universe Einstein envisioned in 1917. And by the time he visited Mount Wilson, Einstein had largely dropped out of the cosmology game anyway. He had made his one towering contribution and had remained largely silent since. To the outside world, however, his endorsement was a momentous event, like a blessing from the pope. (In fact, his word carried much greater weight than the cosmological pronouncements that Pope Pius XII attempted twenty years later.) If Einstein said the universe expands, then it must be so, and the papers reported it accordingly. Hubble was only too happy to face the media and provide any necessary quotes that Einstein, who spoke little English and who by this time had learned how to avoid the press, could not. Einstein's visit to Mount Wilson spread the gospel of the expanding universe and helped secure Hubble's place in the history books.

For Einstein, letting go of the static universe also meant freeing himself from Ernst Mach's theory of inertia, completing his drift away from the empiricist philosophy he once held so dear. Now the question was, what kind of cosmological model could reconcile the new observations with Einstein's old spiritual values? The one thing he was sure about was that it would not contain Lambda, but in every other way he was as willing as ever to fine-tune the universe for maximum beauty. His first pick was an oscillating universe that expands and contracts endlessly, so it could still be immortal. This solution, known as "the Friedmann-Einstein model," had a couple of kinks. First, it implied the current expansion was uncomfortably young. Second, and worse from Einstein's point of view, each rebound appeared to pass through a moment of zero volume and infinite density, a state that he considered nonsensical.

Before starting work on an alternative answer, Einstein took a trip to Caltech, where he crossed paths with his old friend and sparring partner, de Sitter. For anyone interested in cosmology, the stretch from Caltech to Mount Wilson was the place to see and be seen. De Sitter, like Einstein, was pondering how to craft a description of the universe that would take into account the evident motions of the galaxies. The two teamed up on yet another solution, called—surprise—"the Einstein–de Sitter model." This time they aimed for maximal simplicity. Not only did they jettison Lambda, they also found a way to eliminate the overall curvature of spacetime, so that the universe would be flat as a Kansas cornfield. Such a universe wouldn't be static, but it would exhibit a different kind of balance. The gravitational pull of all the matter in the universe would exactly counter the expansion. As a result, the universe is always slowing down but takes an infinite amount of time to brake completely to a halt.

The Einstein–de Sitter universe is still considered one of the simpler and more attractive conceptions of the universe. It surfaced repeatedly in other forms, and it has stuck around as an underpinning of the modern big bang theory. It satisfies the needs of general relativity. It expands forever, so there is no messy end to the universe, and there is no need for Lambda. It does, however, imply the existence of a Lemaître-like beginning of the universe. Einstein and de Sitter simply glossed over this point. As much as Einstein wanted to know the mind of God right now, he had a powerful aversion to any discussion of origins or first causes. Any kind of discontinuity struck him as hideous. This surely explains why he originally called Lemaître's solution "abominable." The vast majority of cosmologists since then have had no qualms about exploring back all the way to the first moments of cosmic time. One might say that Einstein delivered the prophecy of a scientific conception of the universe, but he never entered the promised land.

Einstein tried to present his abandonment of Lambda in the best possible light. He wrote that Hubble's redshifts "can be interpreted . . . as an expansive motion of the system of stars in the large—as required, according to Friedmann, by the field equations of gravitation." The expanding universe is thus "to some extent a confirmation of the theory," he claimed. Note that in the current understanding, galaxies are not zooming through space; it is the space *between* them that expands. Therefore, the reddening of the galaxies described by Hubble really isn't due to the Doppler shift. The change is properly known as a cosmological redshift. As the space between us and a distant galaxy expands, so does the light moving through that space. The stretched light appears shifted to the red end of the spectrum, just like a Doppler shift. But every galaxy, every possible observer, can feel motionless because the movement takes place in space itself. This utterly counterintuitive phenomenon makes sense (to the extent that it can make sense) in the framework of general relativity. Conversely, general relativity leads naturally to this kind of expansion. If only Einstein could have resisted his distaste for beginnings he could have predicted the expanding universe rather than reluctantly embracing it after the miraculous evidence was set before him.

Lambda is visible evidence of a flaw in Einstein's cosmological thinking. Viewed against the backdrop of his legendary brilliance, Lambda has therefore gained considerable notoriety as Einstein's "biggest blunder." This description comes not from Einstein but from the physicist George Gamow, the former student of Friedmann's who was instrumental in developing the big bang model of the origin of the universe. "When I was discussing cosmological problems with Einstein, he remarked that the introduction of the cosmological term was the biggest blunder he ever made in his life," Gamow wrote in his autobiography, *My World Line.* Pundits rarely quote the next sentence, which casts Lambda in a different light: "But this 'blunder,' rejected by Einstein, is still sometimes used by cosmologists even today, and the cosmological constant denoted by the Greek letter Lambda rears its ugly head again and again and again."

Lambda has survived because it is a prime tool for reconciling theory and observation, just as scriptural commentary or midrash does for the traditional religions. Lambda is the leap of faith that reconciles the Word and the world. For instance, there was the pesky problem of the age of the universe. Hubble's recorded velocities, combined with his erroneous estimates of galactic distances, seemed to indicate a universe no more than two billion years old. Lemaître manipulated Lambda so that he could allow as much as hundred billion years of time, which he still worried might be inadequate. These seeming age discrepancies persisted for decades, and Lambda kept popping up as a possible solution. In the 1950s,

Lambda took on a new guise in order to create a model of the universe that expands but has no beginning. In the 1980s, another form of Lambda appeared to explain what happened to the universe during the first 10^{-35} seconds of its existence. Four years ago, Lambda adopted its latest disguise. Now it accounts for the accelerating pace of the big bang. In an irony that Einstein would surely appreciate, it is also invoked these days to make the geometry of space flat—just as space was supposed to be in the Lambda-free Einstein–de Sitter universe.

THE MORE THAT COSMOLOGY became a real theory of the world, the more it mattered to get all the numbers right and make sure they all fit together. As the models grew more detailed and precise, the high priests of cosmology focused on smaller and smaller details of the big creation story. They began to ask not just whether the universe had a beginning, but when, why it started to expand, how quickly it happened, what the exact temperature and density were at a given moment. Each step forward was impelled by a wildly uncertain hypothesis that lingered until shot down and replaced by another one.

So Einstein was wrong—not for invoking Lambda, but for denouncing it. Lambda is often called a "fudge factor," but it is much more than that. It carries the charge of Einstein's cosmic spirit. It stands for the unknown, spiritual element that the scientist desperately hopes will make each cosmological model more beautiful, more complete, more true. It stands for the insane optimism that the world is knowable. It stands for Einstein's inspiring belief that science and reason can edge ever close to true, divine reality, the mystical secrets of the Old One.

CHAPTER SIX

THE ERA WHEN THE UNIVERSE CAME FORTH FROM THE HANDS OF THE CREATOR

On September 29, 1931, the British Association organized a session devoted solely to the topic of "the evolution of the universe." This boisterous sci/religion revival meeting drew a Who's Who in the newly intersecting fields of relativity and astronomy, including Arthur Eddington, Willem de Sitter, and Georges Lemaître. George Gale and John Urani, philosophers of science at the University of Missouri, call this meeting "the birthday of modern cosmology." So many people showed up to hear about the astonishing new theories that the meeting organizers had to open a second hall and amplify the presentations through a set of buzzy loudspeakers. One man was notably absent amid the commotion: the prophet himself, Albert Einstein, who was preoccupied with his search for a unified field theory that would fix quantum theory and banish his nightmare of a God that makes decisions by rolling a pair of dice.

With Einstein out of the game, cosmology erupted into a free-for-all. The great man had shown the way for science to venture into times and dimensions previously considered out of bounds. Now came the question of how to continue down this path, continuing to extend the reach of cosmology while distinguishing it from the old-time religions and philosophies that had resided here before. It was a time like Christianity after Jesus or Islam after Mohammed, as the disciples fought to see who would carry forward the Einsteinian legacy. Which model would advance toward that scientific sublime, the one true, global description of our universe? Lemaître, Eddington, and de Sitter now represented the orthodoxy. They accepted that Hubble's redshifts indicate an expansion of the universe, and they accepted that Einstein's general relativity describes the overall cosmic framework. But there were also heretics in the midst.

The most extreme attack on the expanding universe came from the Swiss-born physicist Fritz Zwicky, Caltech's resident gadfly. He was the perfect person to take on the role of the unbeliever. Fueled equally by inventiveness and indignation, Zwicky relished denouncing his enemies as "spherical bastards"—in other words, no matter how you approach them, they still look like bastards. He was a constant source of brilliant, off-the-wall ideas about midget galaxies and dark matter, many of which fell into obscurity mainly because he so thoroughly alienated his colleagues with knee-jerk reactions against their opinions. Now that everyone had agreed that Hubble's redshifts indicated that galaxies are moving away from us at high velocities, Zwicky naturally decided the whole interpretation had to be wrong. What's more, he believed the whole program of cosmology was off track. A strange observation such as Hubble's redshift law should prompt researchers to consider new scientific principles that they might have missed before, he insisted. Interpreting the redshifts in terms of the well-established descriptions of expanding space guided by general relativity was just a recipe for stagnation. Zwicky was the kind of man who would invent his own faith just to avoid having to go to church.

Working in his accustomed contrarian mode, Zwicky argued that Hubble's redshifts indicated a previously unknown physical process that stretches and reddens light, and he thought he knew what that process was. The cumulative gravitational field of all the mass floating about in space would exert a small drag on light waves, he believed, slowly draining them of energy. In this way, light would grow steadily redder with distance but, as in de Sitter's universe, would not denote actual movement of the galaxies. It was a clever and in some ways constructive proposal. It forced Hubble to be even more careful in his interpretation of the galactic red-shifts. If they were not real motions, then the universe might be static after all, and the new interpretations of general relativity would have to be revised yet again. Physicists could not at the time rule out the possibility of gravitational drag, but neither could they find any evidence for it. Zwicky, who had been unable to get observing time on the one-hundred-inch Hooker telescope and see for himself, railed against his enemies. "Hubble . . . and the sycophants among their young assistants were thus in a position to doctor their observational data, to hide their shortcomings," he fumed.

A more serious cosmological heresy came from the pen of Edward A. Milne, a well-respected stellar astronomer at Oxford University who had previously had little interest in grandiose theories of the universe. His conversion began with, of all things, the letters page of *The Times* of London. In a series of exchanges published during May of 1932, the prominent British astronomer James Jeans spoke out staunchly in defense of Einstein's notion of curved space and insisted that it was the only way to make sense of Hubble's findings. Milne, a small, rigorous man who firmly believed that science should deal only with observable phenomena, worked himself to a slow boil while reading each round of the correspondence. He was appalled by all this abstract talk about the structure of space and dissatisfied with models that offered no explanation of why the universe is expanding rather than contracting, which should be just as mathematically plausible.

One month after the epistolary brawl in *The Times,* Milne published what he considered a much simpler and more sensible explanation for the reddening of distant galaxies. He developed an alternative to general relativity, called "kinematic relativity," which largely preserved classical, Newtonian conceptions of space. In essence, he treated the expansion of the universe the way a laboratory physicist would treat a ball of expanding gas. Milne considered a large group of galaxies scudding about at random. Over time, he pointed out, the fastest-moving ones would naturally migrate to the periphery of the group by virtue of their extreme velocities, while the slowest ones would remain toward the center. Such an arrangement, he noted, would create the illusion of an expanding universe from a vantage within the whole swarming mess. In this way, he could account for Hubble's observations with no curved space and no Lambda. Milne disapproved of applying relativity to the universe as a whole, which he saw as an unjustified extrapolation from known physical laws. "If the curvature of space cannot be determined, if it is essentially unobservable, then it should be rejected," he wrote.

The shock of Milne's comments reverberated through the 1930s. Unlike Zwicky, Milne was no fringe character; for several years his was one of the most widely discussed cosmologies in England, where many of the leading theorists then worked. Moreover, his cleverly constructed argument exposed philosophical issues that certain researchers might have preferred to remain hidden. In Milne's cosmology, galaxies could not be spread evenly through space, as Einstein and his followers had assumed; every observer would see redshifts, but kinematic relativity would not lead to a homogeneous universe. Hubble rejected this almost out of hand. It's not hard to see why: a lumpy universe would wreck his observing program. "The observable region is our sample of the universe. If the sample is fair, its observed characteristics will determine the physical nature of the universe as a whole," he wrote.

But if the cosmic regions accessible to Mount Wilson's one-hundred-inch telescope were not representative of the whole, then that extrapolation would be worthless. Hubble, who tried so hard to cast himself as the incorruptible reporter, could not bear to let go of his sweet faith that his scans of the heavens could reveal the master plan of the universe. Milne also got in a few digs by claiming that his picture of the universe adhered to Einstein's principles, sticking to observable phenomena and understanding them through reason and intuition. The implication was that Milne was not a heretic but in fact a more devout Einsteinian than Einstein.

Five years later, Herbert Dingle, a grand eminence in British astronomy, attacked Milne in an essay entitled "Modern Aristotelianism." Dingle accused Milne of behaving like Aristotle reincarnate, plucking ideas directly from his mind rather than arriving at them inductively from observable truths. The charge, though not unfounded, could have been leveled quite plausibly against every theoretical cosmologist then, and against a fair number of them today as well. During the 1940s, Milne revealed more and more of a neo-Christian religious agenda behind his work. In his view, the only universe that reflects the glory of God is one that begins from a point and expands outward to infinite dimension over infinite time. "In creating an infinite universe, we can say that God has provided himself with the means of exhibiting and practicing his own omnipotence," Milne explained. This belief in the infinite cosmic life span echoed Einstein's initial conception of the universe, not to mention Aristotle's, which placed Milne's cosmology distinctly behind the times. It didn't have any clear grounding in traditional Christian doctrine, but neither did it contribute to the unified picture of the universe growing out of general relativity. Kinematic relativity was a curiosity but, for a few years, at least, a high-profile possibility that scientists had to investigate.

In principle, it should have been possible to distinguish the Milne and Zwicky cosmologies from the orthodox expanding universes promoted by Einstein and his followers. Until the verdict was in, Hubble tried to stay above the fray, continuing to describe the redshifts as "apparent velocities" of galaxies. Seeking guidance from the brain trust at Caltech, Hubble arranged biweekly meetings at his home, where the observers and theorists would sip whiskey, nibble at Grace Hubble's sandwiches, and jot down their latest ideas on a borrowed blackboard. Most fruitfully, Hubble teamed with his friend Richard Tolman, a Caltech physicist well versed in matters of relativity, to find ways to use the images and spectra from Mount Wilson's one-hundred-inch brute to distinguish among the competing cosmologies. His data were not nearly detailed enough to settle the matter, however. If anything, they fit better with Zwicky and Milne than with Lemaître and Einstein, but mostly they proved nothing. Mapping the large-scale distribution of galaxies and pattern of redshifts required even greater light-gathering power. World War II delayed the inauguration of the two-hundred-inch Hale telescope on Mount Palomar until 1948. For the next decade and a half, therefore, Hubble and the other observers had no significant new findings to contribute. Yet on the theoretical side, cosmology progressed at an extraordinary clip during that time, continuing to grow in explanatory power by drawing authority from other corners of science.

The key change was that scientists started thinking much more realistically about the past state of the universe. If galaxies really are moving away from one another—as nearly everyone but Zwicky accepted—then the conditions of the universe must have changed greatly over time. Maybe the powerful telescope atop Mount Wilson could not see those early conditions directly, but there might be ways to search for relics from past eras. Just as archaeologists reconstruct the history of long-lost civilizations from bone fragments and shards of pottery, cosmologists could retrace the history of the universe from elements or radiation created under physical conditions that no longer exist. A few decades earlier, such investigations would have seemed absurd. Everyone still believed, like Aristotle and Newton, that the universe was eternal and unchanging. Individual stars, even galaxies, might evolve and change, but the physical state of the universe was a constant.

Alexander Friedmann took the first step toward tracking the construction of the cosmos back to an initial point, but it was Lemaître who really showed the way into the promised land. His primeval atom hypothesis, introduced in 1931, brushed aside ancient scientific and religious taboos. For the first time, a scientist dared to speculate in a specific, physical way about the origin of the universe. He took into consideration the way that radiation would exert a dominant outward pressure when the universe was very small and even tried to imagine how the rules of quantum physics would play out in such a situation. Equally noteworthy, Lemaître thought seriously about aftereffects of the eruption of the primeval atom that might still be noticeable today: "This highly unstable atom would divide in smaller and smaller atoms by a kind of super-radioactive process. Some remnant of the process might, according to Sir James Jeans's idea, foster the heat of stars until our low atomic number atoms allowed life to be possible." Since scientists at the time did not understand the exact nuclear reactions that powered stars, this was not an implausible idea. Lemaître also mused that some fragments from the primordial cosmic fireworks might still survive. He thought these might explain the existence of cosmic rays, energetic subatomic particles that shower down onto the Earth from space.

Lemaître was no nuclear physicist, so his entire picture of the primeval atom was highly impressionistic. At any rate, scientists had only a vague conception of the workings of the atomic nucleus. The discovery of the neutron, crucial for the development of nuclear theory, occurred in 1932, a year after Lemaître proposed the primeval atom. In general, his ideas received a warmer reception in the popular press than in the scientific literature. Eddington still resisted the idea of "a single winding up at some remote epoch." Hubble considered Lemaître's model "dubious" because it seemed to lead to a universe much smaller and denser than indicated by the photos from Mount Wilson. The search continued for a model that satisfied both the stars and the psyche. Nevertheless, Lemaître's fireworks cosmology got scientists thinking about how to investigate the conditions in the very early universe and connect them to today's observed reality.

Tolman and his friend Howard Robertson, another Caltech theorist, brought greater mathematical rigor to Lemaître's speculations. During the early 1930s, they explored the thermodynamics of the cosmos, analyzing how the background temperature of the universe would have changed over time. Today it is extremely cold in space. In the distant reaches between the stars, temperatures hover just three degrees centigrade above absolute zero. (Absolute zero is the coldest possible temperature, the point at which essentially all molecular motion ceases.) Tolman and Robertson constructed a mathematical model of the universe, filled it with radiation, and watched what happened when they ran the clock backward. If you compress a mass of air—by pumping a bicycle tire, for instance—it grows hot. In these theoretical simulations, the expanding universe behaved much the same way. The cosmic temperature increased in inverse proportion to the average distance between galaxies, the researchers found. When the universe was tiny, a tremendous amount of radiation was crammed into a small space and the universe must have been a hellish place.

Tolman and Robertson considered their work an interesting game, but not necessarily indicative of the real universe. As Tolman wrote in 1934, "We must be specially careful to keep our judgments uninfected by the demands of theology and unswerved by human hopes and fears. The discovery of models, which start expansion from a singular state of zero volume, must not be confused with a proof that the actual universe was created at a finite time in the past." But his warning was instantly self-destructing, like one of those narrated cassette tapes from an old *Mission: Impossible* episode. By pursuing this line of inquiry, Tolman had already displayed a profound hope that mathematical extrapolations from laboratory physics could reveal how the entire universe had evolved. He was grasping for the one dimension that Einstein had sidestepped in his 1917 cosmology, the dimension of time. Still, Tolman wasn't yet prepared to take the plunge and seriously consider what the universe might have been at that "singular state" far in the past. Lemaître, on the other hand, had no trouble speaking eloquently about his ancient primeval atom but lacked the detailed physics knowledge to give more than a hand-waving account of how that atom might have led to the modern universe.

All the elements were out there for sci/religion to take a leap back to the earliest moments of the universe. George Gamow—an energetic and sportive physicist who had a deep grounding in the nascent science of the atomic nucleus—was the first to step forward and seize the opportunity. Many others quickly followed. Born in Odessa in 1904, Gamow decided early in life that traditional religion could not be trusted. After watching Communion in the Russian Orthodox Church, he decided to see for himself whether red wine and bread could transform into the blood and flesh of Jesus. He held a bit of the blessed bread and wine in his mouth, ran home from church, and placed the specimen under the lens of his new toy microscope. It looked identical to an ordinary bread crumb that he had prepared at home earlier for comparison. "I think this was the experiment which made me a scientist," he recalled.

While a student at Petrograd University, located in what is now St. Petersburg, Gamow studied under Alexander Friedmann. The young Gamow fell in love with general relativity and with his professor's visionary notions about cosmology. Other, less pleasant experiences hardened Gamow's distaste for dogma. Russia's new rulers required that university education include training in the Marxist-Leninist philosophy known as dialectical materialism, a confusing hash of ideas based around the notion that progress occurs through the interaction of opposites. Gamow, a natural prankster, didn't take well to being tested on this nonsense and nearly flunked his exam. The Soviet philosophy, he wrote, "played very much the same role as that of Church dogma in the Middle Ages, sometimes assuming grotesque forms." Dialectical materialism was used to justify all manner of arcane beliefs, including, for a time, an official state position disputing the theory of relativity and affirming the existence of the ether. His reaction against the perceived absurdities of Leninist thinking and church doctrine only encouraged Gamow in his playful approach to science.

Not surprisingly, Gamow left Russia as soon as he had the chance, in 1928. He took a fellowship at Cambridge University, where he learned the latest thinking about the atomic nucleus from Ernest Rutherford, a pioneer in the study of radioactivity. In 1934 Gamow settled at George Washington University. During these years, he helped make the first connection between astronomy and nuclear physics by investigating one of astronomy's most visible unsolved questions: How do the stars shine? By this time, most scientists felt the origin of star power must lie at the center of the atom; no other known source of energy could keep a sun shining for billions of years. Another clue came from recent studies of the composition of the stars. Astronomers had long assumed that the stars contained the same mix of elements as the Earth. Around 1925, Cecilia Payne-Gaposchkin, once again hovering at the edges of great discoveries, carefully studied the atmospheres of stars and found them full of hydrogen and helium, the two lightest elements. On the Earth, hydrogen is a secondary constituent mostly locked away in water (the H in H_2O), and helium is virtually nonexistent. The discovery was so shocking that her colleagues refused to believe it until it was repeatedly verified over the next few years. The abundance of hydrogen turned out to be the crucial clue not only for finding the energy source of stars, but also for decoding the first moments of cosmic history.

If stars are full of light elements, Gamow reasoned, then perhaps the heavier elements arose as a result of nuclear reactions in the stellar interiors. Under everyday conditions, such reactions never happen. Positively charged hydrogen nuclei stay away from other hydrogen nuclei, because like charges repel one another. But Gamow recognized that in accordance with quantum physics, particles can briefly bend the rules and overcome that repulsion. If the particles are moving quickly enough—that is, if they are tremendously hot—they can get close enough to fuse together into helium nuclei and release a great deal of energy in the process. Hans Bethe, a German theoretical physicist with similar obsessions, figured out the details of how such nuclear reactions power a star. Gamow was more interested in how the reactions could keep going and synthesize everything in the universe out of two simple particles, the proton (which is the same as a hydrogen nucleus) and its uncharged twin, the neutron. Physicists soon discovered that the reactions in stars could not produce the heaviest elements, but now a fresh idea took root in Gamow's fertile brain. If Tolman and Robertson were correct, the whole universe started out like the center of the sun, only far larger, hotter, and denser. The stars, Gamow suspected, were the bit players. All of the heavier elements—the carbon in our bodies, the iron in our cars, the silicon in the mountains, everything—might have formed in rapid succession in a primordial cosmic fireball.

Gamow had plenty of time to develop these ideas. While many of his colleagues worked on the Manhattan Project, he did not receive clearance and so continued his research. Initially he had pictured a top-down scenario, in which Lemaître's radioactive superatom decayed explosively into lighter elements. But his ideas began to change after the end of World War II, aided by the rapid advances in nuclear physics in the United States. In 1946 he reversed course and imagined a world starting with a thick soup of neutrons—which spontaneously decay into protons and electrons, providing all the necessary ingredients for atoms—merging into heavier elements, bottom-up. Ralph Alpher, then a graduate student at George Washington University and Johns Hopkins, performed the difficult calculations needed to transform the concept into a detailed physical model. This new theory of the early universe mirrored Hubble's discoveries at Mount Wilson. Hubble stretched the reach of science through space; Gamow extended it through time. Both discoveries showcased the growing technological power of the United States, and with it the implied mastery over physics at the largest and smallest dimensions. When he witnessed the first nuclear test, J. Robert Oppenheimer unforgettably quoted from the sacred Hindu text Bhagavad Gita: "I am become death, the destroyer of worlds." Gamow offered an inverse kind of hubris. He promised that the same nuclear theory that obliterated Hiroshima and Nagasaki could read the story of creation back to its very first page. Now Adam could understand how God built the Garden of Eden.

In the beginning, there was light. More specifically, there was a blazingly hot mixture of neutrons and energy, seething and interacting with one another. Gamow called the initial particle soup *ylem,* taken from the Greek word for the unformed state of existence that preceded the creation of the world. In a period of just forty-five minutes, a wave of nuclear cooking swept through the *ylem* and created heavier atomic nuclei. Alpher had calculated the recipe by which an element captures a neutron and moves up one rung in the periodic table, using newly declassified data from the Manhattan Project. The theory worked well, up to a point. It correctly indicated that one-quarter of the mass of the universe should be helium, but it could not account at all for the heavier elements. Nevertheless, this was the first time anyone had dared use modern physics to try to understand a phenomenon that occurred billions of years ago.

The paper describing this innovation was a masterpiece of concise writing: just one page in a 1948 issue of the journal *Physical Review.* It also was a showcase for Gamow's eager sense of humor. He was amused that Alpher's name sounded like alpha, the first letter of the Greek alphabet, while his own resembled gamma, the third. All he needed was a beta so that the byline would read α, β, γ—perfect for a paper about the beginning of the universe. He impetuously added Bethe's name to a paper that he had not written and credited the finished product to Alpher, Bethe, and Gamow. Gamow was so tickled by his little joke that he even attempted to persuade Robert Herman, a physicist at Johns Hopkins who had contributed to the nuclear-cooking concept, to sign onto the paper under the assumed name of Delter.

While calculating the reactions in the *ylem,* Alpher and Herman recognized that the energy from that initial conflagration could not disappear. One of the most stringent principles of modern science is that energy cannot be created or destroyed. So the energy of the initial moment must still exist, though diluted tremendously by the subsequent expansion of the universe. Their 1948 calculations predicted that the fiery early life of the cosmos should leave a background ocean of radiation, a divine light that would keep the universe warmed to a temperature of five degrees centigrade above absolute zero, or −451 degrees Fahrenheit. Such radiation would mostly be in the form of short-wavelength radio waves, or microwaves. This prediction attracted little attention at the time but proved hugely important two decades later. One reason for the lack of interest was that the researchers did not explicitly state that the energy would be in the form of potentially observable microwaves. But the technology of the day probably could not have detected such a feeble signal anyway.

The parallels between the alpha-beta-gamma theory and the

story of Genesis provoked much comment from both the scientific and the religious sides of the fence. In fact, there is no indication that Gamow and Alpher were thinking about the Bible or that they had the slightest interest in using the new sci/religion to prop up the old Judeo-Christian one. Their ideas followed in the tradition of the primeval atom, which grew out of Lemaître's desire to roll back the boundaries of science even if that meant constricting the Christian beliefs that meant so much to him. Gamow didn't even like hearing his model called the big bang. He felt that term emphasized the original instability over the nuclear reactions and subsequent evolution of the universe that formed the heart of his theory. To the general public, however, creation lay at the heart of this new cosmological model. In his dissertation defense, Alpher had remarked that the phase during which the elements formed took about five minutes. This comment was immediately satirized by Herblock, the *Washington Post*'s influential editorial cartoonist, and picked up by a number of newspapers. "There were poignant responses from people who wanted to pray for my soul," Alpher recalled. Gamow sent a copy of his paper to Einstein, who gave it a much warmer welcome. "The idea that the whole explosion process started with a neutron gas seems quite natural," Einstein wrote in response. He approved of the way that Gamow had concocted a single, elegant mechanism to produce all of the different chemical elements and didn't seem too offended that Gamow's work implied a moment of origin for the universe.

In his paper, Gamow did not specify why or how the initial expansion began, but off the record he certainly did think about the matter. His old mentor, Friedmann, had already pointed toward one possible explanation. An expanding universe need not expand forever. If the cumulative mass of the galaxies is great enough, the universe ultimately stops expanding and everything falls back together again. This was the oscillating or "phoenix" universe that Einstein briefly embraced in 1931. Gamow toyed with it again around 1950, giving it the whimsical name "the Big Squeeze." Rather than the endless cycles of Hindu mythology, which Friedmann had envisioned, Gamow pictured one big pileup in which a thin, vast spread of matter and energy fell together, then rebounded to produce the modern universe. He didn't seem overly concerned with figuring out where that earlier universe had come from. Robert Dicke, a leading theorist at Princeton, also investigated oscillating solutions to Einstein's equations, as did some other mainstream thinkers well into the 1960s. They liked the big squeeze for the same reason it appealed to Friedmann: it allowed an expanding universe without demanding a beginning, and it echoed ancient mythologies describing endless cycles of cosmic history.

But the cyclical universe never really caught on. Lemaître rejected the idea of an endless succession of primeval atoms, which fit neither his science nor his philosophy. He believed he had found his own way out by appealing to the laws of quantum physics to obscure the manner in which the universe began. Eddington offered a withering philosophical critique of the cyclical models: "I would feel more content that the Universe should accomplish some great scheme of evolution and, having achieved whatever might be achieved, lapse back into chaotic changelessness, than that its purpose should be banalized by continual repetition." He also disdained the fussy, microscopic nature of Gamow's analysis and preferred to pursue his idiosyncratic, big-picture view of the primeval atom. The theoretical problems were just as bad. Nobody could devise a plausible mechanism that would enable a collapsing universe to bounce and become a new expanding one. Careful calculations showed that radiation could not be destroyed at the end of a cycle, so each universe would be hotter and brighter than the one before; presumably this could not be an endless process after all. Although the oscillating universe never gained a serious scientific following, cosmologists have found ever more creative ways to have an expanding universe that does not require a magic moment when time began.

Gamow's marriage of physics and cosmology established a new tradition of treating the universe as an enormous petri dish. With Alpher's and Herman's help, he assembled a set of equations describing nuclear reactions, plugged in assumptions about the temperature, density, and composition of the early universe, turned the crank, and compared the results to the actual universe. In essence, God's creation was being reduced to an engineering problem. This was also one of the first scientific applications of the newly developed electronic computer, adding to the impersonal connotations of the work. Einstein's original plan had grown much more elaborate. In addition to the equations describing the form of space, now there were equations to trace the history of matter. Yet with every push into the unknown came another mystery, another job for Lambda or one of the cosmologists' other hedges that reconcile beautiful theory with messy reality. In this case, one of the problems was very serious. The nuclear reactions in the *ylem* just plain didn't work as Gamow had hoped. They sputtered out too soon, so that the big bang produced only featherweight elements like helium and lithium. Gold, lead, and the other atomic bruisers had to come from somewhere else.

The hard-charging British astrophysicist Fred Hoyle believed he had already deduced the location of that somewhere else. Hoyle was a determined individualist who proudly maintained his working-class Yorkshire accent during his tenure at Cambridge University. As a child he played with household chemicals, unleashing foul-smelling gases and fabricating batches of gunpowder; he never really outgrew this penchant for causing trouble. Although Hoyle started out working on quantum theory, in the late 1930s he switched directions and threw himself into astronomy, continuing his studies even while developing radar systems during World War II. From early on, Hoyle had no stomach for Gamow's big bang theory of cosmic origin, which he found philosophically objectionable. In fact, it was Hoyle who coined the name big bang, in the course of a series of radio lectures he performed for the BBC. Contrary to many reports, he claims the term was intended not as derisive, just colorful. Still, he had no doubt Gamow was barking up the wrong tree when it came to creating heavy elements. In the late 1930s, Bethe had already showed how nuclear fusion in stars might create elements as heavy as chemical middleweights such as oxygen and nitrogen. It seemed that ordinary stars probably couldn't synthesize anything heavier. But what about extraordinary stars? Hoyle suspected the source of the heavy elements was as plain as the specks of light on the old Mount Wilson photographic plates: supernova explosions, potent stellar detonations that can briefly shine as brilliant as an entire galaxy.

The most common kind of supernova occurs when the interior of a massive star collapses, causing its internal temperatures and pressures to skyrocket. A wave of nuclear burning, just a fraction of an inch thick, tears, through the star, consuming everything it encounters; all the fusion energy that is normally released gradually over a star's lifetime pours out in one burst. Physicists had not yet worked out these details in the 1940s, but Hoyle well understood the potential influence of supernovas on cosmic chemistry. The tremendous density of energy and matter at the center of a collapsing star creates the perfect place to synthesize heavy elements all the way to gold, uranium, and beyond. Supernovas have had a profound influence on the history of science. Bright stellar explosions in 1572 (studied by Tycho Brahe and known as Tycho's Supernova) and 1604 (similarly followed by Johannes Kepler and hence called Kepler's Supernova) proved to Renaissance sky gazers that the heavens are not fixed and unchanging. The Andromeda supernova of 1885 had confused the Great Debate by making it seem as if the spiral nebulae were much closer than their true distances. During the 1950s, supernovas showed Hoyle how to connect nuclear processes with the evolution of stars and the birth of the universe. And in the 1990s, studies of supernovas forced astronomers to revise their ideas of how the universe expands.

The success of Hoyle's supernova analysis cemented his innate skepticism about the big bang. Knowing that some heavy elements could form in exploding stars, he wondered whether all heavy elements might have originated there. This simplifying assumption would eliminate the need for Gamow's hypothetical cosmic egg. And at the time, there were a several good reasons to look for an alternative cosmological model. The most concrete of these was the nagging age paradox. Hubble thought he had obtained a reasonably accurate measure of the rate at which the universe is expanding, and his authority was such that nobody questioned his results for years. If one extrapolates backward from Hubble's reported galactic motions, the cosmos should have been compressed into a white-hot dot less than two billion years ago. But geologists knew the Earth was older than that, and astronomers had strong evidence that stars and galaxies were billions of years older still. The big bang might be a nice creation story, but it didn't seem to leave enough time for the creation part.

Beginning in the 1940s, however, the German-born astronomer Walter Baade gathered startling new evidence that would ease the age problem considerably. Baade, another member of the Mount Wilson brigade, was a skilled and cautious observer. He was considered an enemy alien during World War II, so he was not allowed to participate in military science projects or even to leave the county of Los Angeles. The upside was that he suddenly found it much easier to schedule time on the Hooker telescope. Taking advantage of sensitive new photographic films and exceptionally dark skies—the city lights were extinguished for wartime security—Baade systematically studied the Cepheid variable stars that Hubble and others relied on for determining the scale of the universe. After the war Baade continued this work on the most powerful instrument in the world, the enormous new two-hundred-inch reflector on California's Mount Palomar. Finally, in 1952, he dropped his bombshell. Contrary to what every astronomer believed, there are two kinds of Cepheid variable stars, one much brighter than the other. The ones that Hubble had studied were the brighter variant. Thinking that he was looking at a much fainter population of stars, Hubble had systematically erred and seriously underestimated the distances to his galaxies. Using the recalibrated data, Baade found that all the galaxies were twice as far away, and the universe twice as old, as Hubble originally estimated. Allan Sandage, Hubble's protégé at Mount Wilson, continued this effort and made additional corrections that kept nudging the cosmic age further upward. By the end of the decade, he estimated the universe could be as much as thirteen billion years old. At the same time, astronomers realized that star clusters need not be nearly as old as they had believed. The age problem was a problem no more.

But Hoyle also objected to the big bang on philosophical grounds, and these points of contention never went away. Einstein's original static cosmology had at least offered a single, eternal description of the universe. The expanding universe, on the other hand, had spawned an entire family of interpretations, including de Sitter's, Friedmann's, Lemaître's, Eddington's, and Einstein's own 1932 revision. It was no longer a single doctrine that offered a single picture of the universe. Hoyle found the absence of a unique solution unsatisfying. Models that incorporated some arbitrary amount of Lambda in order to get around the age paradox only made them uglier to his eyes. The big bang implied that the universe evolves over time, which implied that natural laws might also evolve over time. Such evolution would undermine the repeatability of experiments and so undermine the cornerstone of scientific method.

Moreover, the big bang does not specify the initial conditions that gave rise to the universe. Matter and the laws that govern it just came into being. Einstein's cosmological principle required that the universe be homogeneous and uniform in all directions. By every measure, he appeared to be correct. But again the big bang didn't explain why the universe is built that way; it just assumed it happened. To Hoyle, the big bang suggested a horrible limit to human inquiry. It ruled out any exploration of the physical laws and conditions that preceded the initial neutron fireball, leaving science emasculated. "We are forbidden to calculate what happened before a certain moment in time," Hoyle complained. While reading Howard Robertson's massive summary of cosmological thinking, originally published in 1933 but still very relevant, Hoyle found himself deeply dissatisfied. "Has he thrown his net wide enough? Are there any other possibilities?" he wondered. Like Milne before him, Hoyle decided to challenge the new orthodoxy.

To build a better creation mythology, Hoyle teamed up with two newcomers to cosmology. Thomas Gold was a brash young engineer whose only published work was on the physiology of the inner ear. Hermann Bondi was a highly adept mathematician with little grounding in astronomy. Both men were technically enemy aliens—they were Austrian citizens living in England—who became fast friends during their months in an internment camp in Quebec. After their release from internment, they worked on wartime radar research with Hoyle. When World War II ended, Hoyle guided the group toward problems in astronomy, particularly the matter of the expanding universe. Hoyle would ride his bicycle to a house that Bondi and Gold rented, some fifty miles from Cambridge. Then the three men would spend evenings together pondering the workings of the universe, while Hoyle would pace about and ask, "What could the Hubble observation mean? Find out what it could mean!"

After many days like this, Gold hit on the idea of a universe that keeps growing but never changes. Hoyle later claimed, perhaps a bit too colorfully, that the idea was inspired by the 1945 British film *The Dead of Night,* a ghost story whose hallucinatory plot ends up where it began. According to Hoyle's reminiscences, "Tommy Gold was much taken with it and later that evening he remarked, 'How if the universe is constructed like that?' One tends to think of unchanging situations as being necessarily static. What the ghost-story film did sharply for all three of us was to remove this wrong notion." The three researchers proposed a radical alternative to the big bang. In their cosmological model space still expands, ferrying galaxies apart from one another, but new matter spontaneously appears in the gulfs that arise. Over time, that matter would coagulate into baby galaxies that fill the voids. "So we have a situation in which the loss of galaxies, through the expansion of the universe, is compensated by the condensation of new galaxies, and this can continue indefinitely," Hoyle said. *Plus ça change, plus c'est la même chose*—the more things change, the more they stay the same. Such a universe would conform to what Gold called the "perfect cosmological principle." Not only does it look the same in all places, it looks the same at all times.

This "steady state" rival to the big bang arrived in two papers, one by Hoyle, the other by Gold and Bondi, published in the summer of 1948 in *Monthly Notices of the Royal Astronomical Society.* The two camps differed in their approaches: Hoyle's was more mathematical, Gold and Bondi's more conceptual. The latter paper emphasized the key importance of the perfect cosmological principle, which Gold and Bondi saw as an essential reform for the Church of Einstein. "If it does not hold, one's choice of the variability of the physical laws becomes so wide that cosmology is no longer a science," they wrote. Hoyle was less interested in this argument than in the practicalities of grafting a steady, unchanging interpretation onto the existing mathematical descriptions of the expanding universe. But overall, the papers struck very similar themes. The steady state universe required the continuous creation of matter from nothing. In the wide-open and constantly expanding stretches of intergalactic space, new particles would simply pop into existence. The universe is so huge that the creation events need not happen very often. In Hoyle's words, it would require only "one atom every century in a volume equal to the Empire State Building."

Hoyle recognized that many of his colleagues would object that the spontaneous creation violated well-known conservation laws, which state that matter and energy cannot be created or destroyed. "This may seem a strange idea and I agree that it is, but in science it does not matter how strange an idea may seem so long as it works—that is to say, so long as an idea can be expressed in a precise form and so long as its consequences are found to be in agreement with observation. In any case, the whole idea of creation is queer. In the older theories all of the material in the universe is supposed to have appeared at one instant of time, the whole creation process taking the form of one big bang. For myself, I think this idea very much queerer than continuous creation," he explained in one of his BBC presentations. Thus Hoyle cast himself as a purist whose cosmological model flouted conservation laws but strictly followed the most hallowed scientific doctrine, the falsification of theory through observation. The steady state model made testable predictions about how new matter appears. The big bang placed the creation process at some unknowable first moment and so was inherently less scientific in Hoyle's estimation. The steady state also provided a ready explanation for the overall smoothness of the universe: the incessant expansion and regeneration would eventually erase any irregularities, no matter how large. The big bang had to assume smoothness as an inherent condition of the universe. In the 1980s, supporters of the big bang developed a substantially revised version of the theory to address these serious criticisms.

More than Bondi and Gold, Hoyle labored to show how the steady state theory could fit in with general relativity. In order to keep the universe steady and unchanging, he invoked a novel scientific process strongly reminiscent of Lambda. Operating in the tradition of Einstein, Hoyle added an extra term to the relativistic equations to keep everything in proper balance. Hoyle's new term, which he called C for "creation," sat in the same spot in the equations where Einstein had placed Lambda and produced an analogous outward force. "It is this creation that drives the Universe," Hoyle wrote. The difference was that C produced matter, whereas Lambda produced energy to counter the pull of gravity. Again like Lambda, C was a fudge factor, a mysterious element designed to reconcile a beautiful cosmological theory with the observed properties of the universe. Hoyle had criticized the big bang models because they needed a fine-tuning of Lambda in order to get around the age problem, yet now he found himself in the same boat with his rivals. In the search for a comprehensive understanding of the universe, they all invoked a divine hypothesis and prayed that observation would affirm their faith. Through some mathematical juggling, Hoyle argued that the spontaneous creation of matter didn't violate conservation laws after all. In technical terms, he proposed that the C-field had negative energy, so that the increasing strength of the field exactly equaled the mass of the newly created matter—remember that mass and energy are equivalent according to Einstein's equation $E=mc^2$. It still sounded like cheating, but from a formal perspective Hoyle could insist that the scales were balanced.

Steady state cosmology attracted a number of followers, especially in England. William McCrea, a prominent British astronomer and the secretary of the Royal Astronomical Society, actively endorsed the new theory and helped expedite the publication of Hoyle's first paper. McCrea hoped that a complete understanding of the C-term would lead to the unification of quantum physics and relativity, Einstein's unrealized lifelong dream. Hoyle, Gold, and others continued to refine their models. They tried to show that the steady state could account for the formation of galaxies more naturally than the big bang could. From the start, Hoyle considered galaxy formation one of the big bang's greatest weaknesses. "This big bang idea seemed to me unsatisfactory even before detailed examination showed it leads to serious difficulties . . . the really serious difficulty arises when we try to reconcile the idea of an explosion with the requirement that the galaxies have condensed out of diffuse background material," he wrote. He also worked to show that stars alone could account for the origin of all the chemical elements. Gamow's papers on nuclear reactions in the early universe bolstered the credibility of the big bang, but his cosmic fireball could not account for the origin of the heavier elements. Hoyle's studies suggested that massive stars and supernova explosions might be able to do the job. So for a time, the arguments from nuclear physics lent some credibility to the steady state.

Nevertheless, the steady state theory encountered considerable resistance on sci/religious grounds. Many scientists agreed when Gamow derided it as "artificial and unreal." He was more comfortable placing the moment of genesis in the distant past, where it wouldn't interfere with our modern, rational world. Consciously or not, cosmologists had been conditioned by the Bible (and by nearly every other creation mythology) to accept miraculous events that happened at very early times but that no longer operate today. In impish style, Gamow acknowledged the mystical nature of this work but used it to poke fun at Hoyle's complicated mechanisms for synthesizing elements in stars. In a retelling of the story of creation, entitled "New Genesis," Gamow concluded: "And so, with the help of God, Hoyle made heavy elements in this way, but it was so complicated that nowadays neither Hoyle, nor God, nor anybody else can figure out exactly how it was done. Amen." One scientist dispensed with such evasive sarcasm and openly admitted he disliked the steady state model because of its conflict with the doctrines of old-time religion. Milne, Einstein's outspoken critic, was offended by a theory in which the universe came into existence not in a single glorious moment of creation, but in an endless succession of humble field fluctuations. "This is not a Providence that I for one could worship as God; and to do the authors of the theory justice, they do not believe or assert that any transcendental omnipotence is behind the simple acts of creation at all," he wrote in 1952.

At the same time, Alpher and Herman had continued to refine their big bang calculations, aided by a newfangled tool called the computer. By 1953 they claimed they could trace the physical history of the universe back to the first ten-thousandth of the first second of existence. But no matter how much the big bangers couched their theory in abstract terms, talking about particles and fields, they kept running into reminders that the new sci/religion was treading on territory long claimed by old Judeo-Christian belief. Hoyle, a staunch atheist, repeatedly harped on this connection. In his 1950 book, *The Nature of the Universe*—largely a compilation of his BBC presentations—he attacked Christian dogma and linked it with the creation mythology of the big bang.

Pope Pius XII unintentionally strengthened Hoyle's case. In a November 1951 address before the Pontifical Academy of Sciences, Pius XII praised the big bang in great detail. "With that concreteness which is characteristic of physical proofs, it has confirmed the contingency of the universe and also the well-founded deduction as to the epoch when the cosmos came forth from the Hands of the Creator. Hence, creation took place. We say, therefore, there is a Creator. Therefore God exists," he pronounced. He also gushed that "present-day science, with one sweep back across the centuries, has succeeded in bearing witness to the august instant of the primordial *Fiat Lux.*"

Hoyle relished this association between the big bang and the most prominent figure in Catholicism. "I don't take much stock in faith," Hoyle said, tarring the pope and the big bang with the same brush. But Gamow was not the least bit perturbed. Rather, he was so entertained by the papal response that he cited it in a scientific paper published in *Physical Review.* He also sent the pope a copy of one of his recent popular articles on cosmology. Other big bang supporters felt embarrassed by the theory's chummy new association with organized religion. Lemaître vehemently reaffirmed that his cosmology was completely independent of any "metaphysical or religious question" after the pope weighed in. Through a connection at the Vatican Observatory, Lemaître persuaded Pius XII that calling on astronomy to support theology harmed both sides.

Lemaître's concerns were well founded. At the most basic level, the papal address looked like another retreat on the part of old-time religion. Alpher and Hermann had limited the time available for God's act of creation to the first ten-thousandth of a second. Everything after that belonged to science. Even more problematic was the pope's implicit attempt to bolster the credibility of Catholicism by appealing to the authority of science. If science carried weight in this matter, what about all those other matters—the existence of heaven, the resurrection of Christ, transubstantiation—in which it not only provided no support for the Bible, but flat-out contradicted it? The appeal to science leads to deism, not Catholicism. And the evolving truths of science make a very shaky foundation for any ancient religion. Unlike the pope, science never claimed to be infallible. Quite the opposite: the whole source of its power is its fallibility or, more precisely, its falsifiability. Right now cosmology looked good because it implied a single moment of creation, but another observation or a new theory could lead to a different picture of the universe. Sure enough, current ideas about multiple universes undercut the 1950s notion of a single moment of creation. By its very nature, sci/religion does not support the leap of faith that the pope was seeking. He had to emphasize, "*We* say, therefore, there is a creator," because the only true basis for his kind of belief is belief itself. Pius XII never again made a connection between cosmology and biblical creation.

As for that other outspoken authority on science and God, Albert Einstein, he felt ambivalent about the competing cosmological models. To Einstein God was the embodiment of natural law, not the willful Creator of the universe. Not surprisingly, therefore, he always avoided explicit discussions of creation. He thought he had evaded creation entirely in his 1917 paper, of course. When he developed the expanding Einstein–de Sitter cosmology, he never spoke about what happened at the start of the expansion. Even when he discussed cosmology with Gamow, he was interested only in the physics of the early universe, not in the moment when it all went bang. In one of the last interviews of his life, given to Canadian astronomer Alice Vibert Douglas in 1954, Einstein dropped the mask somewhat. He expressed little appreciation for the steady state model, feeling that the universe could have come into existence only through a single formative event. He likewise rejected Milne's kinematic relativity, which he regarded as clever but theoretically flawed. But Einstein did not care for Lemaître's primeval atom, either, and seemed weary of the endless speculation that his 1917 paper had unleashed. "Every man has his own cosmology, and who can say that his own theory is right?" he reflected.

In a sign of just how deeply the new sci/religion had penetrated the popular culture, the cosmology battles took on cold war overtones. When Hoyle assailed the dogmatic nature of the big bang, he associated it not only with Catholicism, but also with nationalism and communism, which, like the big bang, assumed a progressive direction to history. The charge was rather odd considering that Gamow, the high priest of the big bang, detested Soviet ideology and had fled to the West at his first opportunity. Never one to back away from a tussle, Gamow countered with intimations that the steady state was more acceptable to Soviet political theorists, who rejected the notion of a cosmic beginning as the product of idealistic and theistic Western thinking. Indeed, the nominal position of the Communist Party was that the universe is infinite in extent and endless in duration. Pope Pius XII's endorsement merely solidified the perception in the Soviet Union that the big bang model was inherently religious and reactionary. During the 1940s and 1950s, however, the government of the Soviet Union frowned upon any physical theory that attempted to describe the whole universe, dismissing this as bourgeois materialism. As a result, Russian astronomers generally avoided cosmology entirely. The continuous creation process in the steady state theory was, on the whole, just as offensive to Soviet theoreticians as the perceived biblical echoes in the big bang.

While these arguments made for entertaining intellectual theater, the fate of the two competing cosmologies cried out for more of the lifeblood of sci/religion, empirical evidence—and at first, there wasn't a whole lot of it forthcoming. Theory had skipped disconcertingly far ahead of observation and remained there until the mid-1960s. Bit by bit, however, the signs began to look bad for the steady state. Hoyle dismissed the early contrary findings as uncertain and possibly biased, which some of them certainly were. He also had no qualms about modifying the theory. He experimented with a version in which matter and antimatter are created equally, or one in which new matter erupts into existence so energetically that it emits a birth cry of X rays. In later years he considered that the universe might contain large irregularities or might have formed in an ongoing series of little bangs. Hoyle's attitude was always one of exploring all possible explanations. Gold and Bondi, on the other hand, rigorously believed in the primacy of the theory that the universe remained eternally the same. Like Einstein's stationary universe, their steady state model proved an easy target of attack, but it lacked the all-encompassing theoretical power that had made Einstein's cosmology such a revelation to the followers of sci/religion.

If the steady state were correct, at least in its Bondi and Gold version, the cosmos should have always expanded at the same rate. The big bang, in its simplest form, implied a universe forever running down from its initial state. In the 1950s, Humason and Sandage examined galaxies at a wide range of distances to see which description was correct. After much agonizing analysis, they thought they saw signs of cosmic deceleration. If true, it contradicted the most basic assumption of the steady state, although the enormous uncertainties in their data left plenty of room for doubt. (In fact, astronomers in the 1990s found a cosmic acceleration, the exact opposite of what Humason and Sandage claimed in the 1950s.)

Another, far more clear-cut prediction of the steady state model was that all parts of the universe should be largely the same. They should all contain a mix of old galaxies and young ones created from new-formed matter, with no net evolution and a consistent average age throughout. Yet observers noted that no galaxies looked much older than the Milky Way's oldest stars, nor did any look much younger. "It was very clear to me from the beginning that the steady state was wrong. There was never a question in my mind, because having been a student of Baade's . . . it was obvious that all galaxies were the same age as the oldest stellar content. And that could not be in a steady state universe," Sandage explained. Still, this line of evidence against the steady state was inconclusive and somewhat impressionistic. The more devastating blows came from the new field of radio astronomy.

After World War II, the radar and radio technologies developed for military use began to find applications in a peaceful conquest. Astronomers started using the much improved antennas and detectors to explore the universe in a new way, picking out radio emissions coming from various parts of the sky. Some of these sources were obviously familiar objects, such as the sun. Many more were scattered across the sky and were presumed to be galaxies or other distant objects that for reasons unknown uttered a great deal of radio noise. Nobody at the time knew where or what those sources were. Scientists have since identified them as massive galaxies whose emissions are probably powered by gas funneling into a black hole containing as much mass as a billion suns.

In 1954, Martin Ryle at Cambridge University, one of the pioneers of radio astronomy, had a brainstorm. He realized that these radio sources could distinguish between the rival cosmologies. If the steady state were correct, the sources should be distributed equally through space, with the result that faint ones should outnumber bright ones following a simple, geometric pattern. In reality, Ryle soon found there are many more faint sources than one would expect from a uniform distribution, indicating that the radio sources are much more numerous at great distances than nearby. An evolving, big bang universe allows for such changes. The steady state does not. Ryle, who had already feuded with Gold, happily reported that "there seems no way in which the observation can be explained in terms of a steady-state theory." Gold doubted Ryle's results, telling Hoyle, "Don't trust them, there might be lots of errors in this and it can't be taken seriously." But Hoyle couldn't brush aside serious evidence from his university colleague, no matter how much he disliked Ryle's motives in performing these observations. Instead he tinkered with the steady state model, allowing that the universe might be uniform after all. If it contained huge blobs that were all created at the same time, then large regions of galaxies might all have the same age and he could explain why radio galaxies seemed to cluster at large distances. Such a universe could still average out to a "steady state" over very large distances or very long times. To the unconverted, Hoyle's arguments sounded like special pleading. By the early 1960s, Ryle's data were good enough that many astronomers considered them a decisive blow to the steady state cosmology.

But the worst was yet to come for Fred Hoyle and company.

Back in 1948, when Alpher and Herman had speculated that radiation from the fierce heat of the big bang might still ricochet about the cosmos, they didn't regard the calculation as anything more than a curious statistic. They never suggested that anyone look for this relic of creation, nor did Gamow, even though he recalculated the background temperature of the universe several times in the following decade. Starting around 1963, Robert Dicke at Princeton University arrived independently at the same conclusion. Dicke was at the time fascinated by the possibility of an oscillating universe, and he wanted to know how much radiation would be left over from each outward bang. His calculations told him that the radiation, stretched and cooled by billions of years of cosmic expansion, would mostly be in the form of microwaves—short-wavelength radio waves like those used for UHF television broadcasts or, later, for heating food in a microwave oven. He realized that existing microwave detectors were sensitive enough that he could actually test his prediction. James Peebles, a former student of Dicke's who had also settled at Princeton, reworked and refined Dicke's temperature estimate, while two other colleagues went to work building a foot-wide microwave antenna and set it up on a rooftop, hoping to eavesdrop on the echo of the big bang.

While Dicke and Peebles fussed with this experiment, a pair of young radio astronomers at nearby Bell Labs in Holmdel, New Jersey, puzzled over a seemingly unrelated problem. Arno Penzias and Robert Wilson were refurbishing a giant, horn-shaped radio collector originally used to pick up signals bounced off *Echo,* one of the first American satellites in orbit. While calibrating the instrument, the two men were distressed to find some persistent radio noise that they could not eliminate. They took the electronics apart and put them back together. They waited for radiation to die down from a 1962 high-altitude nuclear test. They cleared out a pair of pigeons and some pigeon droppings from inside the antenna, thinking that the body heat from the birds might be the problem. "We were really scratching our heads about what to do," Wilson wrote. Dicke heard of their difficulties and realized that the fault lay not in the telescope, but in the stars. The Bell Labs scientists had a hard time digesting Dicke's explanation of their instrument noise. "Although we were pleased to have some sort of answer, both of us at first felt a little distant from cosmology," Wilson recalled. But Penzias and Wilson had stumbled across the first observational evidence of the big bang. All of Gamow's work on how elements formed in the early universe was an after-the-fact explanation. The microwave noise was something else again. It did what the pope's exhortations could not: it made creation real.

The Princeton and Bell Labs teams published joint papers in the *Astrophysical Journal Letters* the summer of 1965. Penzias and Wilson cautiously reported their discovery of an anomalous microwave source coming from all directions in the sky. It appeared that the heavens were aglow with energy. Dicke and Peebles harbored little doubt that the mystery source was in fact energy left over from the initial expansion of the universe. Lemaître, seventy-one years old and ailing, was comforted by the news that scientists had spotted the presumed afterglow from the fireworks of his primeval atom. Once again a major piece of cosmological research debuted in the daily papers, not in the scientific literature. The news about microwave background appeared on the front page of the May 21, 1965, *New York Times,* affirming the big bang theology squarely to the world. Only then did Penzias and Wilson realize the full magnitude of what they had done.

The amount of energy in Penzias and Wilson's signal was minute. It implied that space has a temperature of about three degrees centigrade above absolute zero, or –453 degrees Fahrenheit. For comparison, Earth's atmosphere would freeze solid at a temperature of about –370 degrees Fahrenheit. But the signal cannot be denied. You see a bit of it in the snow on a TV tuned to an empty UHF channel; you hear it in the static of an FM radio set between stations. The big bang had predicted it. The steady state offered no obvious mechanism for creating such a flood of cold radiation. To this day, nobody has come up with a convincing alternative explanation for the microwave background. "Astronomy leads us to a unique event, a universe which was created out of nothing," Penzias reflected. "Thus the observations of modern science seem to lead to the same conclusions as centuries-old intuition." The steady state model suffered a mortal blow, somewhat reminiscent of what happened to Lambda after Hubble's discovery of the galaxy redshifts. Hoyle fundamentally wanted the same thing Einstein had originally wanted, an eternal and all-embracing description of the universe. He had used a mathematical invention much like Lambda to achieve it. And like Einstein, Hoyle fell before the powerful doctrine of falsification.

The big bang won—for now—and the priests of sci/religion rejoiced that they had advanced another step toward sublime, ultimate truth. With the discovery of the microwave background, cosmologists had at last scored an accurate prediction. This success was to big bang cosmology what the 1919 eclipse expedition was to Einstein's general relativity. Before the event, the theory had beauty and elegance on its side, and that was enough to attract many adherents. Now the theory had observation as well. The adherents became true believers, and the public swayed again before the power of sci/religion. The discovery of the microwave background vindicated Alpher's bold talk about applying numbers and equations to the first five minutes of cosmic history. Time was on the cosmologists' side.

During the 1960s, they continued to probe those first moments in increasing realistic detail. In 1966 Peebles performed a detailed analysis of the primordial nuclear reactions based on the latest data on the density of the universe and a brand-new estimate of the amount of energy in the universe, based directly on the recent measurement of the cosmic microwave background. His model accurately accounted for the observed cosmic abundance of helium, which makes up roughly one-quarter of the matter in every star and nebula. Studies of deuterium, a heavy version of hydrogen, provided further testimony on behalf of the big bang. There's hardly any deuterium in the universe, just one atom for every thirty thousand ordinary hydrogens. But it is peculiar that there is any deuterium at all. Deuterium is a fragile atom that is only consumed, not created, in the nuclear furnaces of stars. It therefore must have originated with the universe itself. A year after Peebles's work, Hoyle and two colleagues studied 144 different possible nuclear reactions. In the end, they quietly concluded that the heat and density of the big bang would have created just the right conditions for synthesizing deuterium, conditions that never occur in the interiors of stars. For a man engaged in a dogged intellectual fight, Hoyle was remarkably gracious to his enemies. Although it did not lead where he had hoped, he praised the knowledge gained by blending cosmology with particle physics. "I think if cosmology had to depend on astronomers, it would be in a much weaker state," he quipped.

By the late 1960s, most astronomers considered the debate closed. The big bang was the official creation mythology of sci/religion, and its details were inscribed in the pages of the *Astrophysical Journal.* The universe began in a ferociously hot cataclysm. That initial mishmash of particles and energy turned to a brew of protons and neutrons bashing into each other, which in turn transformed into a brew of hydrogen, deuterium, helium, and lithium. Stars and galaxies formed from dense regions in the debris. Aeons later, the galaxies are still flying apart, producing the redshifts recorded by Slipher and Hubble. The mix of elements looked right, the estimated age of the universe roughly matched that of the oldest stars, and the microwaves glowed from above like a heavenly blessing on the theory. Peebles reminisced about the mood of those days: "The universe is simple. It's expanding in a computable way—which is an amazing thing. I never would have anticipated it could have worked out so easily, but there it is." Cosmology's unknowns were tumbling like dominoes. Never had the secret of the Old One seemed so close.

Yet the spirit of Lambda persisted. The priests of sci/religion hungered for more complete knowledge and deeper explanations. All of the telescopes and all of the equations still did not indicate why or how the big bang took place. They told nothing of whether the universe would expand forever or someday reverse course and start over again. They did not explain how order emerged from chaos. They did not explain why everything looks so simple: the smooth scattering of galaxies across the sky, the smooth glow of cosmic microwaves, the smooth expansion of space. And they did not speak to the most fundamental puzzles. Why this universe rather than any other? Why is there something rather than nothing?

The big bang needed something more—not just Lambda, but a whole new family of Lambdas.

CHAPTER SEVEN

HISSES FROM THE MICROWAVES

WHAT THE POPE COULD NOT do with his rhetoric, King Carl XVI Gustaf of Sweden accomplished with a handshake, a gold medal, and a piece of paper. On December 10, 1978, the king stood onstage at the Stockholm Concert Hall and handed Arno Penzias and Robert Wilson their shared the Nobel Prize in physics for detecting the pervasive microwave hiss left over from the time of cosmic birth, publicly endorsing the big bang as the modern world's official story of creation. The blessing went both ways. The notoriously conservative Nobel Committee had now officially converted to the faith of sci/religion.

The key elements of Einstein's 1917 prophecy were fulfilled. He had envisioned a universe uniformly populated with galaxies, so that the equations of general relativity could describe all of space. And so it was. He had hoped for a static universe but reconciled himself to an expanding universe in which the density of matter provides just enough gravity to counter the outward motion. And so it could be—the data were not yet conclusive, but such a balance at least seemed possible. He had claimed that physical law holds steady across time. And so it did, by all appearances. Nuclear simulations by Alpher, Herman, and others yielded increasingly detailed and convincing descriptions of the light elements that formed during the first few minutes of the cosmic fireball. Penzias and Wilson's detection of the afterglow of the big bang was the final proof of the power of sci/religion across billions of years of time. The latest cosmological models didn't even need Einstein's rejected Lambda. Astronomers had expanded their inferred age of the universe to about fifteen billion years, leaving plenty of time for the formation of the Earth and the evolution of the oldest known stars.

Yet the high priests of sci/religion were not satisfied. The big philosophical question—Why is the universe the way it is?—continued to nag at them. And as astronomers took better measure of the universe, the question took on increasingly concrete forms. The cosmos appeared to have been built to very exact specifications, and nobody could account for it. Imagine walking into a forest and noticing that the trees are arranged in neat rows and the streams cut geometrically perfect lines. Your map of the region tells you this is virgin forest, where everything should have been shaped only by nature, yet the pattern of the landscape looks very organized. Surely you would be surprised and would want to know the reason for this unexpected arrangement. Cosmologists found themselves in an analogous situation. Once again it was clear that something was missing from the models—some intangible that would make all the pieces fit together.

One sign of trouble, called "the flatness problem," came disguised as good news. A flat universe is one in which the density of matter and the curvature of space-time exactly balance out, so that parallel beams of light neither converge nor diverge as the move. This is the form that Einstein had favored: in the simplest formulations, a flat universe is the same thing as a universe in which the gravitational pull of the galaxies exactly offsets the cosmic expansion. By the late 1960s, astronomers had found that the total mass of the universe is in fact close to that critical density. But Peebles and Dicke, the Princeton University cosmologists who had provided the theoretical interpretation of the ubiquitous microwaves as the afterglow of the big bang, were puzzled. Einstein had never given a reason why the universe should be flat. He just thought it would be simpler and more attractive if God had made the universe that way. If the big bang occurred in a random way, however, the universe presumably could have formed with any density, ranging from zero to nearly infinite. Why should it have almost exactly the critical density, the one that just happens to produce a flat universe that follows the rules of classical Euclidian geometry? Why not a density a million times greater or a trillion times less?

The mystery grew deeper when Peebles and Dicke thought back to the earliest stages of the universe. When the universe was smaller and younger, the effect of any deviation from critical density would have been magnified. For the density to be anywhere near critical today, it had to be astonishingly close at the time of the big bang. To be precise, at the earliest moment that physicists could analyze, the density needed to be within one part in 10^{60} of the critical value. Surely that could not have happened by chance, but no cosmological model could explain why the geometry of the universe is so flat and why its mass so close to the balance point between gravity and expansion, an equilibrium weirdly reminiscent of what Newton had requested of God to keep His universe from tottering. "How did the universe arrange these initial conditions?" Peebles wondered.

As with the rows of trees in the forest, the odds of a beautiful balance in the universe happening through random good luck seemed exceedingly small. Alternately, there might be some as yet undiscovered reason why the universe must have the critical density. Cosmologists thought back to Einstein's philosophical query: "What really interests me is whether God could have created the world any differently; in other words, whether the demand for logical simplicity leaves any freedom at all." As the great telescopes and new satellite observatories revealed more about the kind of universe we live in, the question grew increasingly acute.

In 1969, Charles Misner, a physicist at the University of Maryland, pointed out another kind of unexplained cosmic regularity. He was eager to take cosmology beyond its sci/religious roots based on broad extrapolations from the initial big bang and start exploring the detailed physical structure and evolution of the universe—in essence, moving out of the sketchy story of Genesis into the complicated unfolding of history that followed. During a summer seminar at Cornell University in 1965, Misner heard Peebles talking about the seeming uniformity of the cosmic microwave background, not what one would expect from the chaotic fireworks of the big bang. From the moment they discovered this radiation, before they even understood its cosmological significance, Arno Penzias and Robert Wilson had noticed that the signal in their primitive antenna was roughly the same strength in all directions. "The Earth had made a complete cycle around the sun and nothing had changed in what we were measuring." Wilson reported. Follow-up studies showed that the cosmic microwaves are not just somewhat uniform; they are almost perfectly uniform.

By the late 1960s, better measurements showed the intensity of the microwaves appeared identical in different parts of the sky to within 1 percent, and on fine scales the irregularities were clearly smaller still. Misner sensed trouble. "Things you don't understand can be constant to 10 or 20 percent, but 1 percent requires an explanation," he said. The radiation is so smooth that for years no radio telescope on Earth could find any irregularity at all. In 1989, NASA launched the Cosmic Background Explorer (COBE) satellite to perform the kind of exacting study that could be done only from space. The results from COBE explained why studies from the ground kept turning up empty-handed. The microwave variations are there, but they are minuscule, about one part in one hundred thousand. Such rigorous sameness is practically unheard of in nature, where lumpiness and disorder is the rule. It means that the temperature and density of the early universe, which determined the frequency of the background radiation, must have been nearly identical in all locations.

Again, such consistency might seem like good news. After all, Einstein's cosmological principle assumes that the universe should look generally the same at all places and in all directions. But as Misner reported in 1968, there is something odd about the smoothness of the microwave background. Temperature variations disappear when objects are left in prolonged contact with one another. A hot bowl of soup or a freezing scoop of ice cream will gradually match the temperature of the surrounding room, for instance. Moments after the big bang, however, the universe was expanding at close to the speed of light. Two bits of universe expanding in opposite directions could never have touched each other, nor could they have in any way communicated with each other. Somehow, though, they ended up at almost the exact same temperature. The soup takes hours to reach room temperature. The universe had no time at all—so how did it get so uniform? "I was trying to change the goals of scientific cosmology from describing the universe to explaining it," Misner said in 1990. But the universe was not cooperating.

One could simply give up and assume that whatever process triggered the big bang just happened to produce an exceedingly uniform explosion. Confronted with unexpected order, however, scientists prefer to hypothesize novel mechanisms that might account for it. Eudoxus put together his spheres to reproduce the regular but complicated motions of the planets. Einstein invoked Lambda to produce the kind of balance that occurred in his ideal universe. Now it was becoming clear that Gamow's picture of the big bang needed some additional feature to give it a deeper sci/religious description of how the universe began. Misner called the strange smoothness of the microwave background "the horizon problem," because it seemed as if each part of the universe could look over its horizon, like neighbors peeking over a supposedly private fence, to see what invisible neighboring regions were doing.

The horizon problem leads to yet another puzzle. The smoothness of the cosmic microwave background demonstrates that there were no large deviations from uniformity in the early universe. On the very largest scales, that is still true. Hubble's deep-sky surveys of the 1930s showed that the global distribution of galaxies is uniform, again just as Einstein had assumed. But on every smaller scale the universe is awash with irregularities: the planets, stars, clusters of stars, galaxies, and clusters of galaxies that punctuate the night sky. The cosmos therefore must have started out with some small lumps artfully mixed into the overall smoothness. Once again, the properties at the time of the big bang seem finely tuned. A little less of that primordial lumpiness and the universe today would be nothing but a formless void of cool gas. A little more and all the matter would have collapsed violently in on itself, birthing a hoard of menacing black holes. Either way, we would not be here to contemplate this conundrum.

Even if the models simply assumed a smidgen of lumpiness from the start, for reasons unknown, they still had problems. We know from the microwaves that the early universe was quite uniform even at galaxy-cluster scales. Gravity would amplify any irregularities in the hot gas, but only slowly. Cosmologists had a hard time constructing plausible models that would transform a nearly featureless early universe into the galaxy-riddled one of today. Their models worked a lot better if they added certain kinds of unseen material, or "dark matter," into the mix. Dark matter could provide the extra gravitational pull needed to build galaxies and flatten out the universe, and if its properties were just right, it might not interfere with the uniformity of the microwave background. This mystery material was yet another fudge factor, sort of a Lambda in reverse, designed to amplify rather than counteract the pull of gravity. Each new mystery factor in big bang cosmology pointed to an unfulfilled spiritual element in the new sci/religion.

Dark matter is a generic term referring to objects that emit minimal light relative to their mass. Some of the unseen stuff might be quite unexceptional. Galaxies abound with objects that fit the bill: cool stars that shine only feebly, for instance, as well as radiation-absorbent clouds of gas and dust. Stellar corpses—collapsed remains known as white dwarfs, neutron stars, and, most extreme, black holes—have piled up steadily over the aeons since the big bang. Additional material might be hidden in brown dwarfs, failed stars that never gathered enough mass to start shining. Cosmologists needed a second variety of dark matter, however, one that would give order to the universe without showing up in the microwave background.

Unlike Lambda, dark matter was, perhaps, verifiable: there were strong signs that at least some forms really exist. Fritz Zwicky, the terror of Mount Wilson, first realized that there is more to the cosmos than meets the eye in 1933, just a year after Einstein and de Sitter announced their new, flat cosmology. While he was studying the motions of the various members of the Coma cluster, an extensive gathering of galaxies roughly 350 million light-years away, Zwicky noticed that the galaxies in the cluster zoomed around and past one another much too quickly. Gravity must glue together the Coma cluster, or else its members would have dispersed long ago. But the galaxies in Coma were moving so rapidly that they should in fact be flying apart, assuming the visible stars and gas were the only things there. Zwicky concluded that the cluster must contain an additional component of unseen material that held everything together. In fact, he estimated that 90 percent of the Coma cluster must consist of this invisible stuff.

Poor Zwicky. His colleagues had trained themselves so well to block out his noisy insults and wild proclamations that they missed many of his clever insights as well. They ignored his discovery, and the idea lay dormant until the early 1970s, when Peebles and his equally inquisitive Princeton colleague Jeremiah Ostriker were studying the structures of spiral galaxies and the ways that pairs of galaxies interact with one another. They found that the dynamics of the galaxies made sense only if each system had a thick halo of nonluminous matter extending far beyond the edge of the visible disk.

At the Carnegie Institution of Washington, Vera Rubin, one of the first women to break into the men's club of cosmology, provided even more dramatic evidence that dark matter is real. She started out measuring the ways galaxies cluster together. But driven by curiosity and a frustrating lack of access to the largest telescopes, Rubin switched gears and set out to unravel how spiral galaxies rotated, an arcane subject that had produced lots of theory but few reliable observations. According to the standard assumptions of the time, the outer parts of a galaxy should turn more slowly than the inner parts because they are farther from the center of mass. Rubin found instead that stars maintain their speeds all the way to the very edge of the spiral arms. In a serious of cautious, detailed 1978 articles, she reported that she was seeing direct evidence of the dark matter halos inferred by Peebles and Ostriker. Those peripheral stars were at the edge only of the visible galaxy. A much larger, invisible galaxy provided the gravitational pull needed to keep everything moving apace at the galactic extremities.

All at once, dark matter looked like cosmology's great savior. "The observations fit in so well, since there was already a framework, so some people embraced the observations very enthusiastically," Rubin recalled. Researchers knew that the bright stars and galaxies do not add up to anything close to the critical density of the universe, Einstein's aesthetic goal. But if there were enough dark matter out there, it might be able to close the gap. Much to their delight, the astronomers analyzing images from the giant observatories and orbiting telescopes have since identified an abundance of dark matter everywhere they looked. X-ray telescopes reveal that groups of galaxies are surrounded by huge clouds of hot gas, apparently held together by unseen mass. Dark matter in galaxy clusters betrays itself by the way it bends starlight from more distant objects or the manner in which it draws in hapless neighboring galaxies. On the largest scales, invisible matter seems to outweigh the visible component by about twenty to one.

For the theorists, even a twenty-to-one mix of dark matter was not enough, however. If the density of the universe has the critical value, they needed a hundred-to-one mix; a full 99 percent of the mass must be dark. And most of that dark material could not consist of conventional protons, neutrons, and electrons. The nuclear reactions deduced by Gamow, Hoyle, and their like-minded colleagues set strict limits on the amount of ordinary matter in the universe. Their model of the big bang, which so accurately predicted the composition of the cosmos and extended the reach of sci/religion, works only if the density of ordinary matter is quite low; otherwise the numbers come out all wrong. Also, too much ordinary matter would mess up the smoothness of the microwave background. So most of the dark stuff must consist of some kind of exotic material, perhaps unknown varieties of heavy subatomic particles that do not interact with the particles that make up normal matter. Current physics theories—themselves based on efforts to unify the natural forces, Einstein's old goal—predict that such particles might exist, and cosmologists dream about them happily. For now, though, these "weakly interacting massive particles," whimsically known as "WIMPs," are wholly hypothetical. "Most of the universe must be made from some substance that is not yet identified, and is perhaps not even known," in the words of Alan Guth.

No laboratory experiment has ever produced a dark matter particle; no detector has ever recorded one. In the face of this vacuum of evidence, theorists have nevertheless coined endless names for the inferred particles: axions, photinos, neutralinos. Lecturing on his motivations for the dark matter hunt, Kim Griest of the University of California, San Diego, was quite candid: "As one goes down the list of popular candidates, asking oneself which candidate is the most likely, I have to admit that 'none-of-the-above' comes to mind." Nonetheless, the dark matter search goes on, because without dark matter our picture of the universe makes no sense. Visible matter alone cannot explain the gravitational dynamics of galaxies and clusters of galaxies. Furthermore, a universe overloaded with unseen blobs of ordinary matter would mess up the primordial reactions that Gamow, Hoyle, and company worked out so meticulously. Dark matter particles will be, by definition, difficult to detect. But the faith of the sci/religious holds that these particles will be found. A much-touted dark matter sighting at the University of Rome in 2000 now looks like the product of wishful thinking. One recent experiment gives reason for optimism, however. Physicists working on the enormous underground Super-Kamiokande experiment in Japan have uncovered evidence that neutrinos—wraithlike subatomic entities long considered massless—actually have a small mass. Neutrinos are almost imperceptible and so abundant that they could contribute a significant portion of the dark matter in the universe.

Even the addition of dark matter did not solve the essential mysteries of the big bang; it left a gnawing spiritual hunger. The flatness problem and horizon problem remained unsolved. Nobody knew where cosmic structure comes from. Dicke discussed this unsettled state of affairs in a paper entitled "The Big Bang Cosmology—Enigmas and Nostrums," published as part of a celebration of the centennial of Albert Einstein's birth in 1979. Although he had first warned about the flatness problem a decade earlier, the paper struck a chord within the sci/religious community. The scientific story of creation was incomplete. Einstein had wanted to know whether the laws of physics forced God to build this particular universe. Einstein's disciples still could not provide an answer.

Stephen Hawking turned the question upside down and suggested it might be insoluble: "One possible answer is to say that God chose the initial configuration of the universe for reasons that we cannot hope to understand." He did not pursue this line of thought. Hawking brought it up only to illustrate the path he refused to take. It was a testiment to the authority of sci/religion that such an appeal to an old-style, unknowable God now seemed nearly absurd.

Another, more palatable but nonetheless highly controversial answer comes from logical and philosophical considerations of all possible realities. We could not live in a universe whose laws preclude all the stages of development leading up to the evolution of our kind of carbon-based life, so of course we don't inhabit any of those other universes. In 1974, the British cosmologist Brandon Carter, then a neighbor of Hawking's at Cambridge University, named this somewhat circular argument "the anthropic principle." As Carter put it, "What we can expect to observe must be restricted by the conditions necessary for our presence as observers." This idea has become one of the most debated, reviled, and revered ideas in cosmology.

The anthropic principle has taken on several forms. In its mildest, or weak, version it limits the number of possible physical states that cosmologists consider in their equations to those that could allow humans to exist. A startlingly speculative version proposed by John A. Wheeler of the University of Texas at Austin, the inventive physicist who first described black holes, goes much further. He averred that the time had come "to read the deeper meaning and consequences" from Einstein's cosmology. Wheeler's take on anthropic thinking, known as "the participatory anthropic principle," states that the universe exists only if there is somebody present to observe it. In this sense, the universe must, by definition, have laws and structures that allow sentient life to exist.

Wheeler was not the only serious scientist to follow Carter's lead. Hawking once invoked the anthropic principle to account for the overall smoothness of the universe. It could explain the flatness of the universe and the abundance of dark matter. And it could explain why the laws of physics appear precisely designed to allow this kind of universe, full of atoms and planets. Without all of these things, we never could have gotten to where we are. If the attributes of the universe were slightly different, galaxies would not form, or stars would not shine, or the whole would have collapsed before life started to evolve on Earth. A number of physicists, including Steven Weinberg, another theorist at the University of Texas at Austin, have seriously suggested that there is not one universe but an infinite number of them, each with slightly different natural laws. We merely inhabit the one that is well suited to us, our location inherently selected for us by the anthropic principle. By the late 1980s, the possibility of many universes began to find its way into mainstream cosmological theories.

Yet many scientists view the anthropic principle as little more than an admission of defeat. "It's like throwing up your hands and saying, 'Things are the way they are because otherwise we wouldn't be here to discuss it,'" says Michael Turner of the University of Chicago, a grizzled veteran of theoretical cosmology. "My fear is that we may drift in that direction." The anthropic principle by itself doesn't explain anything; it provides a reason for not needing to explain certain things. It offers a logical reason not to be surprised that the universe seems so finely tuned to our needs. A number of scientists—Hawking again among them—now feel bold enough to propose instead sci/religious explanations of how it all began. These are not testable theories in the conventional sense, at least not yet. But as Hawking says, "There is not much alternative, unless you are going to suppose that God is sending messages into the universe."

So while the anthropic principle could be used to sweep aside the flatness problem, the horizon problem, and the origin of structure in the universe, few cosmologists are willing to cede their hard-won scientific turf to what is essentially a philosophical doctrine. Most consider invoking the anthropic principle only slightly more palatable than invoking the old-style God. They have looked instead for credible hypotheses that could take sci/religion one step closer to explaining why the universe looks the way it does. These new hypotheses heralded the return of Einstein's once deposed Lambda—largely dormant since Baade and Sandage vastly increased the estimated age of the universe during the 1950s—and gave it a totally new look and mission.

Lambda's rehabilitation initially came from the world of the very small, not the world of the very large. As early as 1916, Walter Nernst, a jovial German physical chemist who was friendly with Einstein, speculated that empty space might not be truly empty. According to newly uncovered rules of quantum physics, it could in fact be full of vibrating energy. Such energy could have a profound influence on the fate of the universe because of Einstein's equation $E=mc^2$. Filling space with energy is equivalent to filling it with mass; if there were a large amount of energy hidden within the fabric of space, this energy would produce a huge gravitational field. Wolfgang Pauli, one of the leading quantum theorists, joked in the mid-1920s that the amount of energy predicted by the then-current models would put such a tight squeeze on the universe that it "would not even reach to the moon."

By the late 1940s, a pair of brash New Yorkers—Julian Schwinger at Harvard University and Richard Feynman, then working at Cornell University—gave the quantum world an even weirder spin. In their formulation, the vacuum is a boiling cauldron of activity on the subatomic scale, full of almost-but-not-quite imaginary particles that continuously pop in and out of existence. Again, the implied energy from these particles and fields was potentially overwhelming, but for a while nobody seriously considered their cosmological significance.

Such reluctance was understandable. It was hard enough to believe that such transient, ghostlike particles really exist, much less that they might play a role in our cosmic destiny. But the quantum activity within empty space actually produces some very measurable effects. One of the most dramatic examples of these is the so-called Casimir effect, an attractive force between two closely spaced metal plates, which was predicted by the Dutch physicist Hendrick Casimir in 1948. The narrow gap between the plates limits the number of virtual particles that can appear there; the vast sea of potential particles on the outside therefore pushes the two plates together. The Casimir effect can be detected in the laboratory, and it is just one of many proofs of the reality of virtual particles. Even the fusion reactions in the centers of stars are possible only because particles that should not be able to merge, according to classical physics rules, can squeak through by borrowing energy from the turmoil of empty space. Virtual particles make the sun shine.

It took a while to forge a link between the very smallest and very largest realms of physics research. The man who made the link between virtual particles and cosmic destiny was Yakov Zeldovich—an unfamiliar name to most Americans, but a towering figure in Soviet physics who shared credit with Andrei Sakharov as the father of that nation's hydrogen bomb. After the 1950s, Soviet authorities unshackled him from his military duties, and he chose to focus on an even bigger atomic explosion, the big bang. He saw the task of cosmology as both ridiculously ambitious and deeply romantic. "We are in a difficult position, knowing that we study directly a small part of the Universe as a whole, and knowing that yet unknown physics of very high energy is involved. One needs courage. But one needs also delicateness and precision. I would call it the Leo Tolstoy principle: a detailed courageous study of his own heart and mind helped him to understand other hearts and minds—that of Anna Karenina, that of a horse," Zeldovich said in an address delivered shortly before his death in 1987. For security reasons, Zeldovich was not allowed to leave the Soviet bloc, but his intellectual fame was sufficient to draw the world's top physics talent, including Stephen Hawking, to his Moscow lair.

Zeldovich was a small, intense man given to frequent and sudden brainstorms. He developed a highly influential model of how structure formed in the universe. He helped show how black holes could function as the energy sources of erupting galaxies. And in a 1967 paper he recognized the universal implications of the energy hidden within the fabric of space in the form of virtual particles. That energy, rather than bringing the heavens crashing down, could give the vacuum an elastic, springy quality, as if space were exerting an outward pressure. A large volume of space would contain more energy, and so exert more pressure, than a small volume. On a local scale, the effect of the vacuum energy might be nearly imperceptible, but over huge distances the squirming of space could create a repulsive, antigravity effect.

In other words, Zeldovich realized that the energy hidden within vacuum would exactly mimic the properties of Lambda. Thus Zeldovich took a big step toward attaining one of Einstein's greatest sci/religious visions, finding a unifying link between the large world of general relativity and the tiny world of the quantum. This reincarnation of Einstein's Lambda is not quite as miraculous as it sounds. In his 1917 cosmology paper, Einstein had tried to put together the broadest possible framework for how the universe might be constructed. He therefore envisioned two general kinds of influences that could affect the dynamics of the cosmos as a whole. Gravity bends space from without, causing collapse; Lambda unbends space from within, causing expansion. The reason vacuum energy fits the description of the original Lambda is that Einstein had already left a large hole in his equations for some generic phenomenon that injects energy, and hence repulsive pressure, into the fabric of space. In fact, Fred Hoyle had already exploited this hole in much the same way with his "C-field," the driving force behind the now discredited steady state theory. Even earlier, Lemaître had recognized that the Lambda in his cosmology equations resembled an energy packed into the vacuum.

The big problem came when people tried to calculate the value of Lambda associated with the vacuum energy. Making such a calculation requires a detailed understanding of all the quantum processes that produce virtual particles. Nobody has yet achieved that level of enlightenment, but physicists allowed some simplifying assumptions that reduced the problem from impossible to agonizingly difficult. After all the number crunching came the embarrassing result. Zeldovich assumed that most of the vacuum energies cancelled out, and still he got a value for Lambda that is 100 million times too large. Using the current standard quantum models, the discrepancy is far worse. The energy density of the vacuum, according to theory, should be about 120 orders of magnitude greater than is observed. A hundred and twenty orders of magnitude is 1 followed by 120 zeroes. If Lambda were actually that large, everything—you, the seat you're sitting on, the pages you're now reading—would fly apart in far less than the blink of an eye. Some missing process evidently trims Lambda to a manageable size.

But once Zeldovich had shown that Lambda might result from known physical processes, it suddenly had new credibility as a part of the sci/religious cannon. Some researchers sought to explain why the vacuum energy is so small, while others sought to understand how it might influence the evolution of the cosmos. Edward Tryon, then just beginning his academic career at Columbia University, tried something much more radical. Starting in the late 1960s, he explored the possibility that the vacuum energy could explain not just the expansion of the universe, in its guise as Lambda, but where it came from in the first place. In essence, he was suggesting that the rules of physics could absorb the old-time role of God as creator in addition to God as rule-maker. Even Einstein had never attempted anything so outrageous.

Tryon thought about the particles that constantly appear out of nowhere in the vacuum. Normally they disappear before they can have any permanent existence. But what if just once the process got out of hand? Maybe the entire universe emerged from a vacuum fluctuation that, instead of immediately collapsing back in on itself, expanded outward in a rush of spontaneously generated matter and radiation. If so, our universe is a fluke, a random twitch of physics. Einstein, who hated physical explanations that depended on a roll of the dice, would have been aghast. *Physical Review Letters* rejected a paper describing Tryon's wildly fanciful proposal, and many scientists took it as a joke. But in 1973, the prestigious journal *Nature* published Tryon's paper, "Is the Universe a Vacuum Fluctuation?" "I offer the modest proposal that our Universe is simply one of those things which happen from time to time," he wrote.

As the 1970s progressed, physicists began to feel increasingly at home in cosmology, and vice versa. As this new kind scientific unity took hold—convergence between very large and very small—Tryon's ideas started not to look so silly after all. The impetus for this interdisciplinary dabbling once again came from a project started by Einstein. His cosmic religious faith told him that nature should be harmonious, whereas the physical laws described in textbooks sounded like cacophony. So starting in the early 1920s, Einstein attempted to show that gravity and electromagnetism, two very different kinds of natural forces, are in fact two aspects of a single fundamental force that can be described by a single set of equations. This mystical project occupied and confounded him for the last three decades of his life.

Twenty years after Einstein's death, three physicists partially vindicated his effort. Using elaborate mathematical arguments, Sheldon Glashow, Abdus Salam, and Steven Weinberg exposed an underlying commonality between electromagnetism and another force, the weak nuclear force that governs radioactive decay. The physicists showed that these two forces should in fact appear identical at high temperatures. In the early universe, when everything was extremely hot, the two forces were one and the same. All of a sudden, particle physicists wanted to know more about the physical conditions immediately after the big bang. At the same time, cosmologists had a new tool for exploring how the universe evolved into its present state.

All of these mind-spinning ideas—vacuum energy, Lambda, unified forces, matter emerging from nowhere—came dancing together in the head of Alan Guth during the late 1970s. At the time, Guth was an eager physics grad student at the Massachusetts Institute of Technology, working under Weinberg, who indoctrinated him into the latest "grand unified theories," which attempted to create an even more general physics theory that combined all forces except for gravity into a unified whole. Guth was no astronomer; he stumbled into cosmology only in a roundabout way. He was actually trying to fix a vexing glitch in these unified physics models. Theory predicted that the universe should be filled with junk particles, called "magnetic monopoles," that should have formed almost immediately after the big bang. But the doctrine of falsification through observation soon cut that theory down: experiment after experiment revealed not a trace of the expected particles.

Looking for a way out, Guth played around with different descriptions of the vacuum energy in the early universe, which depended on some highly esoteric guesses about how the laws of physics behave at temperatures and pressures that lie far beyond the reach of experiment. Why not? After all, he was a youthful researcher working in a speculative realm where there were no absolute answers. At the end of 1979 he found, to his delight, that there is indeed an alternative account of the first moments of existence that gets rid of the unwanted monopoles. Basically, he did it by resurrecting Lambda, but in a much more potent form than anything Einstein envisioned.

In Guth's revised scenario, the universe passed through a momentary phase just after the first moment of existence when a tremendous amount of energy is trapped in the vacuum. That energy acted like a supercharged Lambda, a repulsive force that caused the infant universe to expand at breakneck speed. This energetic phase would have lasted just a moment, but it was a very eventful moment. During the brief episode—about 10^{-35} second, so when Guth said brief, boy, he really meant it—the universe grew by a factor of 10^{30}. The human mind cannot really fathom such minute and enormous numbers. If you expanded an atom by a factor of 10^{30}, it would dwarf our galaxy. The visible universe, which before the growth process was far smaller even than an atom, emerged about the size of a very large beach ball. Perhaps inspired by the staggering jumps in the American consumer price index during the late 1970s, Guth called this period of runaway cosmic expansion "inflation."

Guth had explored inflation as a way to sweep away the unwanted monopoles, by expanding the universe so much they were diluted to undetectability. But he soon recognized that inflation had much bigger implications for cosmology. It resurrected that strange model of empty, exponentially expanding space that Willem de Sitter had introduced as "solution B" in 1917. Inflation resurrected Lambda and again gave it a central role as shaper of the universe. It implied that we see only one tiny corner of a much vaster universe, which might be a trillion trillion times bigger than what is visible to us. And above all, as Guth quickly discovered, it handily solved several of the big problems looming over the big bang.

A year earlier, Guth had heard a lecture by Dicke in which he had discussed the flatness problem. Now Guth had a brainstorm. Return once more to the rubber-sheet analogy of the universe. In the first instant after the big bang, that sheet could have had any arbitrary shape. But if you kept stretching the sheet, it would begin to look flat no matter how it started out. Guth realized that the inflation process would do the same thing to the shape of spacetime. No matter what the initial conditions of the big bang, inflation would produce a flat universe with a mass very close to the critical density. In a fit of giddy enthusiasm, he jotted down this idea in his notebook under the heading SPECTACULAR REALIZATION. This page now sits like a holy relic, on display behind glass at the Adler Planetarium in Chicago.

Soon after this insight, Guth had another flash of inspiration during lunch at the Stanford Linear Accelerator Center (invariably referred to as SLAC), where he was working at the time. "December 1979 was my lucky month," Guth writes. "A few weeks after the invention of inflation, I stumbled upon another key piece of evidence to support it." Some theorists were explaining one of the other key cosmological conundrums: the smoothness of the cosmic microwave background, that much debated horizon problem. Inflation could provide a cure here, too, Guth realized. The universe could have started out extremely irregular, with hot and cold spots in any random pattern. Once inflation set in, however, any one spot would grow so large that it would fill our entire visible universe. It would not matter if our particular spot were a cold one or a warm one. The important point is that the temperature would be everywhere the same, just as astronomers observe when they study the cosmic microwave background. Any primordial lumpiness in the big bang would likewise get smeared out in the course of the mad stretching caused by inflation. Interestingly, the old steady state model of cosmology offered the same basic solution to the horizon problem, only it assumed that the smoothing took place over billions of years of normal expansion, not an inflationary frenzy lasting a tiny fraction of a second.

Too much smoothess just brings us back to the question of how galaxies could have formed. But further theoretical explorations revealed a plausible explanation for the origin of structure in the universe. The answer lies once again in the quantum cornucopia of the vacuum. Recall that empty space is constantly pulsing with energy and matter. Normally these fluctuations are invisibly small, comparable in scale to subatomic particles. During inflation, however, the vacuum blips would have been tremendously magnified and stretched along with everything else. Stephen Hawking calculated in 1982 that inflation would expand quantum fluctuations to about the right size and density needed to form galaxy clusters. He thus linked Guth's ideas to something concrete and observable. If he were correct, subatomic physics operating at the smallest scales we can study is directly responsible for the largest structures we can see—another piece of beautiful sci/religious harmony.

Inflation was an instant smash in the world of cosmology. Suddenly, everyone wanted to know more about the shaggy-haired upstart from SLAC who had cured the big bang of its ills. There was only one little problem: As originally formulated by Guth, inflation didn't work. He quickly ran into problems when he explored the details of how the extra energy is first trapped and then released from the vacuum.

In a remarkable instance of parallel thinking that recalled the independent discovery of expanding cosmologies by Friedmann and Lemaître in the 1920s, Andrei Linde at the Lebedev Physical Institute in Moscow had developed a theory of inflation almost identical to Guth's during the late 1970s. He failed to see the merit in the idea, however, and in a fit of despair decided "there was no reason to publish such garbage." Then in 1979, Alexei Starobinsky at the Landau Institute for Theoretical Physics, also in Moscow, hit on a more realistic version of the theory, and Linde got excited again. "For two years, it remained the main topic of conversation at all conferences on cosmology in the Soviet Union," Linde explains. In the West, however, this work was unknown. By 1981 Linde had heard of Guth's difficulties and devised an improved version of inflation that circumvented the problems. Two physicists at the University of Pennsylvania, Andreas Albrecht and Paul Steinhardt, arrived independently at the same fix. The new version of inflation differed in many technical respects, but its outline was unchanged: It started with a brief early era during which Lambda was enormous, causing the universe to swell exponentially, and ended with a return to conditions that put the universe back on the conventional big bang track.

Within a few years, inflation entered the sci/religious mainstream, completing the marriage of cosmology and theoretical physics that began in the 1940s when Gamow, Alpher, Hermann, Hoyle, and others had used nuclear physics models as a tool to test their notions about the big bang. Guth and Linde were now blending ideas from quantum physics, unified field theory, and observational cosmology to push the authority of science even closer to the first moment of existence. "The big bang says nothing about the bang itself—it really starts just after the bang. Inflation attempts to fill in the gaps, providing a prehistory that gives a possible explanation for the initial conditions that the big bang just assumed," Guth says. Once again, Lambda was helping to uncover the secrets of the Old One.

Unfortunately, inflation theory did not offer much for the observers to test in the 1980s. There were a few ways to see if Guth's version of Lambda was any more credible than Einstein's, but they were difficult and likely to turn out inconclusive. For instance, the simplest version of inflation predicts that the density of the universe should be almost exactly the critical value. But measuring the cosmic density was no mean feat, especially considering that most of it seems to exist in an invisible form. Inflation also predicted a distinctive distribution of lumpiness in the universe, the fluctuations that seeded the formation of galaxy clusters. Maps of cosmic structure were not up to the task. Guth and Linde needed inflation to score some obvious empirical successes before it could properly join the expanding universe and the primal fireball in the canon of sci/religion. For the moment, inflation had the status of a powerful but unverified prophesy, similar to Einstein's cosmological model of 1917.

Emboldened by half a century of remarkable cosmological advances, many theorists were willing to ignore the lack of evidence for inflation, because they saw the theory as the only way to explain why the universe looks the way it does without a distasteful detour back into old-time religion or an unholy alliance with the anthropic principle. Veteran cosmologists regularly call for caution until sci/religion has brought more evidence to bear on the issue. "Inflation is a new idea, very attractive, I don't know any alternative to it, but I don't know if I should believe it," says Peebles. But other members of the congregation are eager to proceed toward revelation. "I do not think we'll be able to disprove the theory of inflation," says Linde, who lauds it as the "only explanation" for the odd flatness and smoothners of the universe. Inflation shoved aside the nasty anthropic principle, which is no longer needed to explain the fine-tuning of the universe, and required no special appeals to God. Best of all, inflation gave new respectability to the kind of starry-eyed speculation pioneered by Edward Tryon when he suggested that our universe popped out of the nothingness of the quantum void.

One of the most exciting and baffling aspects of inflation is that it gives a plausible explanation of where all the matter in the universe came from. During inflation, the expansion was driven by the tremendous amount of energy trapped in the vacuum. At the end of the inflationary era, an abrupt change occurs. The amount of energy trapped within the vacuum plummets, and this energy appears in tangible form as heat. So much heat is released at this point that particles literally boil out of the vacuum. During this episode, still within the first moment of the first second of existence, all the mass of the universe pops out of nowhere. Inflation is the reason the big bang went bang.

Any sensible person would ask how something could be created from nothing. Einstein's famous $E=mc^2$, which states that matter can transform into energy and vice versa, provides a partial answer. But if the matter came from energy, the energy still had to come from somewhere. What powered the big bang? Guth's cunning reply is that the mutual gravitational attraction among all the particles in the universe gives rise to an enormous amount of negative, or potential, energy. A vase sitting on a high shelf has negative energy relative to the floor. If it falls off the shelf, the energy is liberated and the vase gains enough velocity to shatter itself when it hits the ground. Likewise, Guth notes, the universe abounds with negative gravitational energy that would be liberated if everything were to fall together. This energy is so great that it could exactly equal the mass of the universe.

Guth, Linde, and the many other inflation theorists argue that the negative energy created as particles expanded away from each other in the early universe exactly balanced the positive energy needed to create that matter. Matter in motion creates the negative energy that allows the creation of more matter. Voilà! Something from nothing. As Guth likes to say, "The universe is the ultimate free lunch." (Actually, Guth notes that a tiny universe already exists when inflation kicks in, so the lunch is not entirely free. He corrects himself and declares that "the universe is a very cheap lunch.") It sounds like unbelievable sleight of hand, but other scientists take the idea as a serious solution to the problem of our origin. In the inflationary view, once inflation ended and all the particles popped into existence, the universe would look just like the hot big bang that cosmologists had envisioned since the time of Gamow's landmark 1948 paper. There would be no obvious sign of where the matter came from or why it was now so uniform.

For all that, inflation still leaves some of the most fundamental questions hanging. It doesn't explain why the strength of gravity, the electron structure of the carbon atom, and countless other details are exactly what they need them to be for people to exist. It doesn't explain why we inhabit a universe whose properties allowed inflation to occur in the first place.

"Why was it necessary for God to work to such precision?" Linde wondered. He has provided his own answer with an elaboration of the idea, called "chaotic inflation." Linde envisions that the universe arose among the random field fluctuations in a preexisting universe. In one spot, the state of the quantum fields was just right to touch off an episode of inflation, and there our universe was born. But this was far from a onetime occurrence. As Linde pictures it, new regions of space are constantly budding off old ones, inflating, experiencing their own big bangs, and evolving into entire separate universes that could each have their own physical laws. A heroic kind of super-Lambda takes on the role of the Creator, keeping the process going endlessly.

"When I invented this theory, I was so excited I stopped everything," he says. Chaotic inflation seemed to him an amazingly general and economical way to account for the nature of our cosmos. If there are, as he believes, an infinite number of universes, then we do not have to wonder about what happened before the big bang or what lies beyond the visible edge of the cosmos. Our universe is just a speck in an eternal, limitless "multiverse." "Here the universe produces itself in all possible forms," he says. This idea echoes the endless cycles of renewal and decline postulated by the pre-Christian Greeks and contains elements of the oscillating or "phoenix" universe that once appealed to both Einstein and Gamow. But the self-reproducing universe has a theological beanty that neither of these possessed. If Linde's interpretation is on target, then the anthropic principle starts to make more sense. Most of the universes may well be hostile to life; many will be short-lived; maybe every other one is sterile. But with an endless number to choose from, it is not surprising that at least one of them turned out to be just right for us. "Anything in physics that isn't forbidden is compulsory," Guth says. "Once you accept eternal inflation, anything that can happen will happen."

Lee Smolin at Pennsylvania State University, another kindred spirit, views this multiverse as analogous to the natural variations that allow biological evolution to occur. Most mutations are harmful, but every once in a while a beneficial one occurs. When that happens, natural selection quickly ensures that the organism with the slight innate advantage will come to dominate the population. In an analogous way, Smolin pictures a precession of universes, each having slightly different physical rules, that sprout whenever a star collapses into a black hole. Most of these are sterile places, but a universe whose laws allow the formation of stars soon begets other universes full of stars, planets, and perhaps life. These cosmic cradles quickly fill with living organisms. What we call "the cosmos" is one of these celestial oases.

The existence of a boundary in time—a moment of creation—always irked Einstein. Linde and Smolin cleverly dodge the temporal bullet by adding near-endless cycles of creation that push the moment of origin almost unthinkably far away. The process could have initiated with a quantum fluctuation that conjured up a true something-from-nothing universe, but still that fluctuation had to arise under the guidance of some preexisting set of natural laws. "So where do the laws of physics come from?" Guth asks. "Science has no answer at this time." The sci/religious faithful find themselves at an impasse reminiscent of that of Saint Augustine, who believed that God created everything, even time itself. What then did God do before the existence of time? "Some people say that before He made Heaven and Earth, God prepared Gehenna for those who have the hardihood to inquire into such high matters," Augustine wrote evasively in his *Confessions.* Hawking accepted Augustine's implied challenge and helped develop a complicated theory, the "no-boundary proposal," in which time can wrap on itself so that it can be finite and unbounded, equivalent to the finite but unbounded space of Einstein's 1917 cosmology. "The universe would be completely self-contained and not affected by anything outside itself. It would neither be created nor destroyed. It would just BE," he proudly writes.

Even as cosmologists plunged bravely toward Augustine's hell, their theories still left an awkward gap between the divine era of Lambda-driven creation and the more mundane world of today. All of the grandiose interpretations of inflation focus on events that occurred billions of years ago or in alternate universes cut off from our own. According to the usual interpretation, inflation ended when the cosmos was just 10^{-30} second old. From then on, the energy drained from the vacuum and the big bang went about its business calmly, without any influence from Lambda. Almost everyone believed that inflation was a phenomenon of the distant past. "The universe is not expanding in an inflationary way today," Hawking asserts in *A Brief History of Time.* Likewise, Guth doubted that Lambda played an important role in the modern universe. He reiterated that the leading unified theories of particle physics predict a value of Lambda that is 120 orders of magnitude larger than the limits set by observations of the expansion of the universe. Clearly, something is wrong with the theories: either the expected vacuum energy does not exist or some process cancels it out. Either way, the most natural assumption is that the value of Lambda is zero. It seemed incredible, to Guth and to everyone else, that the universe would find a way to eliminate 99.999999999999999999999999999999999999999999999999999999999999999999999999999999999999999999999999999999999999999999999 percent of the vacuum energy but leave exactly enough to influence the expansion of the universe without messing up the everyday physics of our lives.

Now that the idea of Lambda was once more a part of mainstream cosmology, however, some other theorists continued to play with it. Perhaps some residual Lambda survived after inflation ended. A dab of energy might yet lurk within the vacuum. In the search for the sci/religious divine, Einstein's old cosmological constant still made for an awfully convenient fudge factor to bring harmony to the universe. The age problem still vexed astronomers: the oldest stars appeared to be older than the estimated age of the universe. As Georges Lemaître had discovered decades earlier, Lambda could clear up that problem handily. More fundamentally, efforts to measure the density of the universe kept coming up short. If some of the cosmic mass existed in the form of vacuum energy, that added component could increase the tally and reach the magic "critical" density that produces a flat universe. But Lambda was still viewed as something of an embarrassment. It had fooled Einstein, allowing him to believe in a static cosmos. What researcher could hope to succeed where the greatest mind in physics had failed?

This was where cosmology remained stuck for nearly two decades. Theory had gone about as far as it could go. In the back-and-forth process that propels science onward, the next move belonged to the observers. They made the most of the opportunity.

CHAPTER EIGHT

THE ANGEL OF DARK ENERGY

SAUL PERLMUTTER darts around his modest office at Lawrence Berkeley Laboratory, a cluster of drab-looking buildings nestled in the Berkeley hills above the University of California campus. Surrounded by institutional yellow walls and gray steel bookshelves, he riffles through a stack of journal articles and computer printouts, searching for the right one. With his edgy movements, shaggy hair, and self-deprecating, Woody Allen–ish gestures, he could be mistaken for a computer programmer—and he does in fact spend a lot of time at the keyboard. Then he picks out the key article, "Measurements of Omega and Lambda from 42 High-Redshift Supernovae," and it becomes clear that the papers and computers and the entire California landscape are only a minuscule, superficial part of who he is. Perlmutter led the movement that revised the big bang and reinvigorated cosmology with a fresh dose of mysticism. He is one of the new high priests of sci/religion.

Perlmutter works in the tradition of Edwin Hubble and his self-proclaimed observational approach to cosmology. Like his illustrious predecessor, Perlmutter is driven by a low-key but irrepressible desire to see all the cosmos has to offer. "It goes back to childhood. I've always been interested in the most fundamental questions. I wanted to know the fundamental rules that make things look the way they do. This time around, I wanted to do something experimental—I wanted to see something about the world around me," he says. Much has changed since the 1920s. Silicon light detectors have replaced glass photographic plates, and rumpled T-shirts have taken the place of jackets and ties during observing sessions, but the drill remains the same. You peer deep into the heavens, scan the countless specks of light, and search for the deeper levels of reality that nobody else has seen before. Once again this effort has given credibility to some of the most startling aspects of Einsteinian prophecy.

Along with a large group of collaborators, Perlmutter set out to determine how far he could go in penetrating what Hubble called "the dim boundary—the utmost limits of our telescopes." Another team, under the direction of the soft-spoken Brian Schmidt, embarked on a similar quest. Both Perlmutter and Schmidt were in their twenties when they began. They were the children of the sci/religious revolution; they had grown up never knowing any doubt that cosmology could provide an all-encompassing view of the cosmos. But from inside the Church of Einstein they watched as theorists compiled a list of neotheological questions. Is inflation the correct description of the big bang? Is space full of invisible energy? Where did our universe come from? Perlmutter and Schmidt wanted to drag those questions back into the observational realm.

Both men had the same basic goal. They wanted to uncover two of the most sought numbers in cosmology, the rate at which the universe is expanding and the deceleration parameter—how the pace of expansion is changing over time. These two values contain some of the most basic information about the nature of the universe we live in. Together they indicate when the universe began and how much matter it contains, pulling the galaxies together and slowing down their outward movement. And there was another, tantalizing possibility. If Lambda exists—if Einstein's 1917 conjecture was correct—then that should show up, too.

Perlmutter initially sought cosmic truth through studies of subatomic particles. By 1987 he was fed up with complicated physics experiments that would take years to provide any meaningful answers and set out in search of a different way to study the world. "It looked like astrophysics was going to let me get at fundamental problems," he recalls, so he struck up a collaboration with his Berkeley colleague Carl Pennypacker to measure the cosmic expansion rate. The duo expanded into a team as they picked up graduate students and colleagues to help with the effort. Eventually it mutated into the Supernova Cosmology Project, with Perlmutter in charge of an ever-changing team lineup.

Schmidt initially had a more concrete goal. He wanted to understand the mechanics of a supernova. When a star obliterates itself this way, for a few weeks it shines with more than a billion times the luminosity of the sun. In the process, as Fred Hoyle discovered in the 1950s, it spews a cloud of heavy elements that seed the universe with the heavy elements that make up planets and people. "I like them as physical objects. What are they doing inside? Why are they so bright?" Schmidt wondered. But like Perlmutter, he recognized that supernovas are also powerful tools for exploring cosmology's great spiritual questions. In 1994 Schmidt joined the hunt for enlightenment with some guidance from his mentor, the supernova guru Robert Kirshner of Harvard University. Schmidt called his effort the High-Z Supernova Search. (Z is the term cosmologists use to denote redshift, so high-Z refers to extremely distant objects whose light appears greatly reddened by the expansion of the universe.) In addition to a late start, Schmidt had to compensate for a geographical disability. In 1995 he took a position at Mount Stromlo and Sliding Springs Observatories in New South Wales, Australia (now the Research School of Astronomy and Astrophysics), which placed him a dozen time zones away from many of his colleagues.

Others had set down this path before. Hubble dreamed of mapping and analyzing the farthest cosmic depths using the two-hundred-inch Hale telescope, the "Big Eye" that opened on Mount Palomar in 1948. He began with a burst of late-life optimism. In 1951 he laid down a grand agenda: "We shall turn to the great problems of the universe with new confidence. Observational results can be stated positively, with limits of uncertainties evaluated accurately. Then theory after theory can be eliminated. . . . And possibly, just possibly, we may be able to identify, in the shortened array, the specific type that must include the universe we inhabit." His plan was to continue, on a vastly larger scope, the research that had led him to discover the apparent expansion of the universe in 1929. He would search for Cepheid variable stars in distant galaxies, this time knowing that such stars come in two varieties. He would measure how quickly they pulsed in order to determine their true luminosity and, hence, their distance. Then he would combine those measurements with Humason's redshift data to see how quickly the universe was running down. He even held out a vague hope of peering all the way to the "observational horizon" of the universe. Two years later Hubble died of a cerebral thrombosis, his program barely begun.

Allan Sandage and Hubble's other successors carried on this enormous undertaking. Even the Hale telescope could produce detailed images of Cepheids only in a handful of fairly nearby galaxies. Beyond that it was hopeless; individual stars were just too dim. Hubble had used cunning tricks to extend his reach. He looked at the brightest stars in galaxies, assuming they must all be similarly luminous; farther out, he looked at the brightest galaxies in clusters of galaxies. In this way, he attempted to add rung after rung to his cosmic distance ladder. But the uncertainties in each step of extrapolation were so great that Sandage and his rivals were still bitterly debating the correct rate of cosmic expansion four decades later. Humason, meanwhile, found that the background glow of the atmosphere obscured the redshifts of the faintest galaxies. "Well, there is apparently no horizon, at least as far as the two-hundred-inch goes," he said with a sigh. Einstein and his disciples could describe the whole universe in their equations, but the observers were struggling to test the truthfulness of those mystical visions.

Back in 1938, Walter Baade at Mount Wilson had suggested another way to take the measure of the universe. Frustrated by the same limitations that had held back Hubble, Baade thought about tracking supernovas rather than Cepheid variable stars. Supernovas are so much more luminous that they can be seen clearly regardless of how far away they are. If all supernovas were essentially the same, they could be used as "standard candles" to reckon the correct distances to galaxies billions of light-years away. But as Baade and others soon learned, the universe does not yield its secrets so easily. Some exploding stars are at least five times as luminous as others. Without understanding the nature of those variations, a naive cosmologist might arrive at distance measurements that were off by a factor of two—far too crude for the delicate business of mapping the exact physical parameters of the expanding universe.

In 1941 the German astronomer Rudolph Minkowski recognized that supernovas fall into two broad categories: Type I, which do not appear to contain hydrogen; and Type II, which do. During the 1950s, Hoyle worked out the basic theory of supernovas as thermonuclear detonations, titanic relatives of the hydrogen bomb. Their diversity made it clear that there is more than one way for supernova explosions to occur, however, and distinguishing among them was not easy. Starting in the 1960s, astronomers began to recognize that the two types of supernovas are completely different kinds of objects. Type II supernovas are the death throes of massive stars that have exhausted all the nuclear fuel in their cores. With no energy blazing out from the center, the star collapses, creating so much heat and pressure that all of the star's outer layers undergo nuclear burning all at once. But at least some Type I supernovas occur through a different process. When a middleweight star like the sun grows old, it ends up as a stellar remnant called a white dwarf. Usually the story ends there. But if the white dwarf has a companion star, it can grab material from its partner and keep growing more massive. Eventually it hits a critical point at which gravity can no longer support all that bulk, and the star caves in on itself—then producing a nuclear blast similar to the one from a Type II supernova.

By the late 1970s, astronomers knew enough about the different types of supernovas that a number of eager cosmologists thought it was time to revisit the possibility of using the exploding stars to gauge distances and put the increasingly exotic cosmological theories to the test. The idea took off in 1985 when Sandage, working with Gustav Tammann at the University of Basel, uncovered a subdivision within the Type I supernovas. One specific kind of supernova, referred to as Type Ia, always seems to blow up the same way. These objects alone—not all the other, similar-looking Type I stars—are exploding white dwarfs. Fortunately, these explosions have a distinctive look that makes them easy to identify, and as luck would have it, they are also the most luminous variety. Type Ia supernovas are so potent, they are easily visible all the way across the universe, perfect for charting the future course of our cosmos.

Perlmutter got excited. "It became clear that Type 1a supernovas were the really useful ones, the ones that could provide a standard candle in a way that we had never thought we had before. I drew light curves and they were amazingly close—they all fell on top of each other. It was clear there was some physics making that happen. And the spectra all looked so similar. It looked like we had a tool we could use for distance measurements and the Hubble constant," he says. At the time, he was still an aspiring particle physicist rattling around at Lawrence Berkeley Lab looking for a juicy research project. Nothing really captivated him until he heard the holy song of the supernovas and saw the light of sci/religion. Here was a chance to tackle the biggest physics problem of all, the origin and fate of the cosmos. In essence, Perlmutter was rediscovering the motivations behind Einstein's 1917 cosmology: to find the one set of laws that explains the universe. He learned to express this argument forcefully over the years in order to convince Lawrence Berkeley Lab that its sponsor, the Department of Energy, really should be spending its dollars on a cosmology program.

In 1988 Perlmutter and Pennypacker split off from the rest of their research group and started hunting for supernovas among nearby galaxies, with "nearby" meaning that they were no more than a few hundred million light-years distant. And so the Supernova Cosmology Project was born. The first of many daunting tasks Perlmutter faced was simply finding the supernovas. Historically, astronomers have considered these stellar detonations extremely rare events. On the local scale, that is certainly true. The last visible supernova in our galaxy was the bright star recorded by Johannes Kepler in 1604, five years before Galileo turned his first telescope skyward. In any one galaxy a Type Ia supernova lights up roughly once every three hundred years. On the large scale, however, the numbers were on Perlmutter's side. There are so many galaxies in the universe—about one hundred billion of them, according to recent estimates from the Hubble Space Telescope—that a supernova visible to today's largest telescopes appears every few seconds. Based on statistics alone, Perlmutter would have plenty of events to study. The difficulty lay in locating a single faint blip of light from among the sky's countless specks and then gathering enough of its feeble glimmering to reconstruct the tale of the recently deceased star.

Just a few years earlier, such a project would have seemed laughable. This time the stumbling block was not telescope power, it was manpower. Nobody could have mustered the time or the staff to comb through all the images on photographic plates looking for a telltale flash, schedule follow-up observations around the world, and analyze with maniacal precision the rising and falling brightness of the supernova. The only reason Perlmutter could now dream of tackling such a challenge was because of some stunning developments in electronics technology, most notably more powerful computers and digital, silicon light detectors called "charge-coupled devices," or CCDs. "We often treat the history of science as a back-and-forth between theorists and observers, but an equal part of that back-and-forth is the technology. It's more like a three-legged stool, with the technology allowing you to do things you couldn't do before. Einstein gave us the conceptual tools to ask these questions, but we didn't really have the technology tools until five or ten years ago for the kind of stuff I'm doing. Now the experimental results we're getting are going to push cosmology forward again," he says. Hubble hacked away at the secrets of the universe with the glass and steel of the Hooker telescope. Cosmology had advanced from the industrial era to the information era, and the key tool was the silicon chip.

CCDs aided the hunt for supernovas in part because they are much more efficient than photographic film, so they can squeeze more performance out of a telescope. More important, they keep an electronic record of every iota of light they receive. This digital tally makes it a snap to single out one exploding star in a field full of galaxies. So Perlmutter could start by creating a reference image of a patch of sky. He would then look again a few weeks later and cross his fingers that bad weather would not ruin the scheduled observing session. Next he would subtract the first image from the second. The digital sky images consist entirely of binary ones and zeroes, so the subtraction should leave nothing but background noise if everything stayed the same. Anything that wasn't there the first time around would pop out immediately. If one star exploded and brightened, it would now be hard to miss.

It sounds simple enough. In practice, nobody had yet made the system work effectively. Perlmutter found himself spending long hours writing software to combine, clean up, and analyze the CCD images. "A lot of times you think, Boy, you're spending your whole life on this stupid computer," he says with an enthusiastic laugh. The technology of the late 1980s, before the World Wide Web had even been invented, was barely adequate for the task. An effective supernova search requires a wide-field camera that can look at a lot of galaxies all at once, maximizing the chance that an explosion will appear in any particular image, but early CCDs were quite small. Commonly available computers barely had enough power to process all the data collected by the telescopes. Network links were essential for shuttling information from observatory to observatory, but the Internet was still an exotic and cumbersome device. There's a reason it was used back then primarily by scientists and technofreaks.

Slowly the situation improved. CCDs got bigger, computers got faster, and Perlmutter began to develop more effective software. "One of the reasons the supernova studies seemed plausible to us was because the technology had advanced so far. We were some of the first people to develop CCD controllers. The computers went through such a change during that same period. When I started, people were using PDP-10s—they didn't have much memory. VAX computers came along and suddenly the programming became much easier. When we started doing the high-Z [large redshift] search, the computers were not quite up to it—but within a year they were. And then networks came up to speed. I was calling people at NASA, asking to borrow lines to Australia by hook and by crook. At every stage of the game, we didn't quite have the technology we needed, but that meant we were ready when the next step came along," he explains. There were political hurdles as well.

Large observatories schedule telescope time only for people who have plausible observing programs, and nobody was sure that the Supernova Cosmology Project could really achieve its ambitious goals. "It was a chicken and egg problem, so you haven't yet found the supernovas, so it's hard to ask for all the time that you want," Perlmutter says.

The first big break came in 1992, when the Supernova Cosmology Project bagged its first high-redshift supernova using a new, large CCD detector on the four-meter (thirteen-foot) telescope at Kitt Peak National Observatory in Arizona. Although this single observation was too sketchy to have much scientific merit, it proved that the team was on the right track. Two years later Perlmutter bagged seven supernovas during observing sessions at Cerro Tololo Observatory in Chile, which has become something of a mecca for supernova hunters. The technique was working. But amid all the excitement, the sci/religious faithful began to disagree on how to interpret the heavenly messages.

One of the fellow seekers joining Perlmutter in Chile was Brian Schmidt, who started to feel uneasy about the work. "We were not terribly happy with the way they were analyzing the data at the time—partly it was out of ignorance, partly it was real scientific differences. We thought it would be good for us to do our own, independent analysis of the data the way we wanted to do it," Schmidt says. When he returned to the United States, he met with Kirshner and suggested that the time was ripe to launch a separate supernova search. Kirshner was skeptical at first—people had found supernovas before but not been able to squeeze useful information from them. "Yes, we could do it better, but could we *do* it?" he asked. Schmidt convinced him that they could. Together with a number of other supernova experts in their circle, the twenty-seven-year-old Schmidt began the High-Z Supernova Search and set off on his own road to enlightenment.

Perlmutter had the advantage of several years of software development. He had also been toughened by his years in the LBL physics group, where people were familiar with fussy, highly technical research projects. But Schmidt had two potent factors acting in his favor: a group of colleagues intimately familiar with hunting for supernovas and a tremendous ability to rise to the challenge. He quickly called his former competitors and organized them into a loose confederation. Then he sat down and hammered away at the same programming problems that had so consumed Perlmutter. Kirshner becomes instantly boastful when he reflects on his disciple's achievements: "Saul's group worked for six years on software, and Brian said, 'I could do that in a month.' And he did." Perlmutter had laboriously created much of his software from scratch. Schmidt, drawing on his greater familiarity with the astronomical world, swiftly cobbled together existing computer programs into a workable, if less elegant, solution.

Both teams needed the computer to help them with the insanely complicated task of understanding the supernovas. Silicon processors allowed them to go where Hubble could not. Baade's old headaches still plagued them, however. As difficult as it was to find the Type Ia explosions, it was even harder to understand what they were doing. The scientists wanted to know both how bright each supernova appeared and how bright it truly was, but the cosmos is not a straightforward place. The expansion of the universe reddened the light of each supernova by a slightly different amount. That color change would affect how the CCD detectors measured the light and needed to be accounted for. A scattering of dust between the supernova and us could also change its color and would reduce its apparent brightness; these factors demanded due consideration. Worst of all, preliminary supernova surveys conducted during the late 1980s and early 1990s showed that Type Ia supernovas are not all identical after all. Astronomers' standard candles came in slightly different wattages.

These problems came to a head in 1991, when astronomers observed two relatively nearby Type Ia supernovas of startlingly different luminosities. The sharp disparity suggested that observers might have to find another way to measure distances. On closer examination, however, the situation was not so grim. Some exploding stars brighten and fade significantly faster than others. Those early surveys uncovered enough Type Ia supernovas in relatively nearby galaxies that a pattern began to emerge: sluggish supernovas are consistently brighter at their peaks than fleeting ones. In these proximate galaxies, there were other ways to measure distances and evaluate the supernovas. Scientists had no such luxury for remote objects, where supernovas were the only predictable beacons bright enough to be visible. But the correlation between speed and luminosity was so tight that the shape of a supernova's light curve—a plot of its changing brightness over time—could very accurately determine the intrinsic brilliance of the detonated star regardless of its location.

Enter Adam Riess at Harvard, another member of Kirshner's flock. Drawing on this information, he helped devise a statistical technique for eliminating the variations among Type Ia supernovas. A little later, Perlmutter came up with his own, more geometric solution in which he stretched the light curves to correct for the differences. Either way, the teams now claim they can figure out the inherent brightness of the explosions to within about 10 percent, an astonishing level of accuracy in the world of cosmology. The supernovas had vindicated themselves. They seem to be almost perfect distance markers. For the first time, scientists had a tape measure large enough for the entire universe—an observational tool that could match the far-reaching spiritual power of Einstein's equations. Now it would be possible to apply the rule of falsification and see what kind of universe God had created. The two teams were off and running.

To be accurate, it was more like the two teams were off and chasing each other through knee-deep molasses. Hunting supernovas calls for a singular mix of frantic activity and almost limitless patience. It begins in a frenzy of administrative activity, securing time on a large telescope just after a new moon, when the sky is dark, and about three weeks later when moonlight again is not a problem. The four-meter (thirteen-foot-wide) telescope at Cerro Tololo, sitting under ink black skies fifteen thousand feet above the Chilean desert, is equipped with a special detector, the Big Throughput Camera, that can see an especially large swath of sky all at once. This camera can snap an image of five thousand galaxies in just ten minutes. Both Perlmutter's and Schmidt's teams depend on this versatile instrument. Once they have secured two images of the same areas of the sky, they must make sure the two views are properly aligned, then account for changes in atmospheric clarity, then eliminate the many flickering objects that are not supernovas, such as variable stars, quasars, and rogue asteroids. During each observing session, they might look at fifty or one hundred parts of the sky, covering hundreds of thousands of galaxies.

If a blip of light looks promising, a new round of work begins. Then the scientists make their pilgrimage to the huge Keck I telescope atop Mauna Kea, on the Big Island of Hawaii. Thirty-six perfectly polished and aluminized glass hexagons work in unison to form Keck's thirty-three-foot-wide cyclops eye. Keck gathers enough light from the supernova suspect to spread the beam into a spectrum, which contains a wealth of data about the star's composition. That information makes it possible to distinguish Type Ia supernovas from the other varieties of exploding stars. Once the astronomers make a positive ID, the real frenzy begins as they scramble to keep the star under nearly constant surveillance in order to produce a sufficiently accurate light curve. Supernovas don't sleep, and neither do the people who study them. For the Supernova Cosmology Project, Perlmutter juggles observing time on telescopes around the globe: the Cerro Tololo Inter-American Observatory 4-meter telescope, William Herschel telescope, Wisconsin-Indiana-Yale-NOAO telescope, European Southern Observatory 3.6-meter telescope, Nordic Optical 2.5-meter telescope, Keck telescopes, and the Hubble Space Telescope. In all, the scientists need to track each supernova for forty to sixty days to get an accurate read. After that comes more analysis to correct for intergalactic dust and other possible sources of error. Final analysis can take a year or more, until the exploded star has faded from view, when it is possible to get a clean view of the galaxy where it lived and died. Meanwhile, each team felt the other breathing down its neck.

Perlmutter's Supernova Cosmology Group had established an early lead. Despite an abundance of talent, Schmidt's High-Z Supernova Search got off to a rocky start. Just five months after he got the project going, Schmidt made that move to the Mount Stromlo observatory in Australia, about as far as possible from the colleagues whose efforts needed close coordination. "I had just had a child, I had just written software that had never been used before, and I was attempting to look for supernovas and debug the software across twelve time zones between Chile and Australia. It was nearly a disaster," he says. Then in 1995, shortly after getting started, Schmidt found his first cosmologically significant supernova, and it was clear his efforts were not in vain. The exploding star was dimmer than expected, suggesting that the universe was up to something odd, but Schmidt was too careful to draw any major conclusions from a single data point. The High-Z team hunkered down to gather more data.

By the end of 1996, Perlmutter had collected twenty-four distant supernovas and completed analyzing seven of them. These explosions lay four billion to seven billion light-years from the earth, or as much as halfway to the visible edge of the cosmos. Now came the moment of truth. Nobody had ever tested the big bang this way before. All those elaborate theories about dark matter and Lambda were about to meet the sharp sword of observational falsification.

Soon, it seemed, humans would know the fate of their universe. If the cosmic expansion were slowing down, as everyone expected, then the light from the most faraway supernovas should be a little brighter than it would be if there were no deceleration, because the diminishing rate of expansion would not have carried them as far away from us as it would have otherwise. This was, in fact, what his preliminary results seemed to indicate. The Supernova Cosmology Project also had some bad news for fans of Lambda. Those first seven supernovas showed no clear sign of any hidden springiness in space. Lambda was probably small, if it existed at all.

Within a year, however, the picture reversed. As Perlmutter worked through more of the observations, the average explosion was looking fainter and fainter. In other words, the universe was not slowing down as much as his team had thought, which meant that the amount of matter in the universe had to be quite modest. Thinking back, Perlmutter now brushes aside his original conclusions as the product of rough, preliminary data. Kirshner bitterly recalls, however, that he stood by those findings until additional observations forced him to change his tune. At any rate, the real answer emerged over the course of 1997.

A follower of Hubble's puritanical style of sci/religious faith, Perlmutter believes that cosmic truth will reveal itself if you just pursue it hard enough. "If the data are telling you something you're not happy with, you have to go with the data, but if the data are telling you something surprising, you have to assume there's something wrong with it. We all didn't believe it at first—assumed it would go away. Little by little we all got convinced. We never sat back and changed our minds. We'd just say, 'I don't believe it, do you believe it?' Well, must be this thing wrong," he explains. But with each new observation, the data were growing more insistent, yet the apparent rate of deceleration—and hence the inferred cosmic density—was looking smaller and smaller.

"It was getting ridiculously low," Perlmutter says. He started saying that if the numbers fell any further, he'd have to conclude that the density of the universe is zero. "I guess we're not here," he joked nervously. Then the numbers fell even more, and Perlmutter found himself looking at minus signs. The supernovas all looked too dim. They were farther away than their redshifts seemed to suggest, evidence that the space between them and us has been expanding at an accelerating rate. In other words, the universe is not slowing down at all. It is speeding up.

Even a completely empty universe could not speed up unless some outward pressure were acting on it. Reluctantly, Perlmutter and his teammates turned to Einstein's old mystical hedge, Lambda, to make sense of the universe. "I told my review committees that we could go after the vacuum energy of the universe [that is, Lambda], which is clearly a fundamental physics question. But I didn't seriously think I was going to find it," Perlmutter says. Like almost all of his colleagues, he assumed Lambda had vanished after the era of inflation. Wanted or not, the intangible found its way into his research. An accelerating universe means that "empty" space must be full of energy.

Late in the year, members of the Supernova Cosmology Project started showing their startling results at scientific gatherings. But the real bombshell came on January 8, 1998, when Perlmutter presented his analysis of forty distant supernovas at a press conference at the American Astronomical Society meeting—a high-profile summit of professional astronomers, which is always extremely well attended by the media. The next day, headlines like SCIENTISTS SEE COSMIC GROWTH SPURT popped up in the papers, Saul Perlmutter was a science celebrity at the ripe age of thirty-eight, and cosmologists were another step closer to reconstructing the history and fate of the universe.

Schmidt's team members had also been making significant progress. During 1997, they picked apart the light from more supernovas and kept finding that, as with the 1995 explosion, the stars were all peculiarly dim. In the autumn, Adam Riess had finished studying the light curves and sent his findings to Schmidt, who was stunned by what he saw. It was obvious to him that the big bang was accelerating. At the time, though, Perlmutter's only public results stated the exact opposite. "I was very concerned. Why were we getting completely different numbers from Saul's?" Schmidt wondered. He kept the findings quiet while he checked and rechecked the results. Little did he know that Perlmutter's latest supernovas were now telling the exact same tale.

While Perlmutter was painting a picture of the runaway universe to an enthusiastic press corps, the High-Z Supernova Search was keeping a low profile. The group did not yet have a completed paper making its own case for cosmic acceleration. When Schmidt's results went public at a cosmology conference a month later, most of the media treated it as confirmation of a story they had already reported. Perlmutter casually reinforces that view: "Our competing group came in with a result just a few months later that agreed well with ours." Schmidt bristles at being seen as an also-ran and focuses on when the groups were doing their research rather than when they announced their results. "To be fair, we came out with the answer at the same time. We ended up converging on the same answer, though their view has ended up shifting over time," he says. Kirshner sounds even more bitter, accusing Perlmutter of misleading people about the chronology of events: "For some reason, he's eager to establish it wasn't a dead heat, but it was." At the end of 1998, *Science* magazine recognized both teams when it named the discovery of the accelerating universe its "breakthrough of the year."

Given the intense competition between the Supernova Cosmology Project and the High-Z Supernova Search, astronomers were particularly impressed that they had arrived independently at almost the exact same result. This is one of the key ways in which sci/religion trumps the old-time religions. It airs its differences openly to help sort out the biases, hopes, and dumb mistakes that influence its discoveries. In its incremental and sometimes blundering way, sci/religion follows the light that leads toward ultimate, unobtainable truth.

The two teams developed distinctive ways to interpret their galactic images, evaluate the light curves, and translate them into distances; they also mostly worked with different data, although they did share two supernovas in their early results. Yet they came up with almost identical statistics on the density of the universe. According to the supernovas, all the matter in the universe adds up to one-quarter to one-third of the critical density, the amount of matter necessary to halt the motion of the big bang. That in itself is a significant finding, because it means that the expansion will continue unchecked forever unless the physical state of the universe changes drastically. The new results largely eliminate the hope of a "big crunch" that will start a new cycle of existence. The supernova data also jibe well with the increasingly refined methods for weighing the universe, such as counting the numbers of galaxy clusters and measuring hot gas around the clusters held in place by their gravity. These techniques showed that all the matter, visible and dark, comes to about one-third of the critical amount, closely matching the supernova's story.

The true stunner was that Perlmutter and Schmidt also agreed that we live in a runaway universe where Lambda, not matter, controls our destiny. "That seemed to me pretty shocking," Guth says. Nobody from Einstein on down had ever seriously proposed such a possibility. Einstein surely would have been baffled by the return of his greatest blunder, which he had created to bring order to the universe and abandoned to bring back common sense. This discovery also continued Einstein's mystical quest by showing that space is an equal partner with matter, and both are accessible to human investigation. Although Lambda is energy, not matter, it has an equivalent mass and density (another consequence of Einstein's $E=mc^2$). The Supernova Cosmology Project and the High-Z Supernova Search processed their numbers and found that their results fit very nicely with a universe in which one-third of the density is matter and two-thirds is an unknown form of energy, possibly Lambda. The total density would then be exactly the critical value, just as Guth and Linde had proposed two decades earlier in the inflationary theory of the big bang.

The amount of energy contained within the hypothetical Lambda is overwhelming on cosmic scales but minuscule by human standards. If you scooped up a block of empty space 250,000 miles on a side—about the distance from the earth to the moon—you'd find just about one pound of energy inside, assuming you could find a magical technique for weighing it. In that same box of space, you'd find roughly half a pound of ordinary matter, mostly hydrogen atoms. The universe is very nearly empty. The battle between Lambda and matter is thus a fight between almost nothing and even closer to nothing. But the conflict has an important implication for the history of the universe. If the universe is accelerating now, its expansion was slower in the past and its age is greater than one would expect simply by extrapolating backward from the present motions. With Lambda in the equation, Perlmutter places the cosmic age at fourteen billion to fifteen billion years, a nice, comfortable range that allows plenty of time for the first stars to form and that settles the "age paradox" that has bugged cosmologists ever since Hubble.

Theorists immediately went to work trying to understand this new development. Vacuum energy could come in many forms. Michael Turner of the University of Chicago, a thoughtful and outspoken cosmologist, coined the generic term "dark energy" because, he says, "'funny energy' didn't sound serious enough." The accelerating universe demanded respect. At the American Astronomical Society meeting, Perlmutter was still stunned by significance of the supernovas. "Even as we were talking about it, we hadn't realized the full enormity of what we were saying," he explains. Schmidt was similarly befuddled. And Kirshner recalls talking about cosmic acceleration with his former students in the High-Z team: "I thought it was a terrible idea. I didn't think it was right and didn't think it was plausible. But in the end, it became clear the problem wasn't with the data reduction—the problem was with the universe."

Or was it? All these grandiose conclusions—the runaway universe, expansion without end, and a world where Lambda or some other form of dark energy, not matter, dictates our ultimate fate—were hanging on observations of a handful of supernovas. In Perlmutter's set of forty-two supernovas, the average one was too faint by 0.3 magnitude, or about 25 percent. Considering past errors in cosmological measurements, that sounds like an awfully small effect from which to draw giant inferences. Hubble's first distance estimates, after all, were nearly ten times smaller than the modern values. Both the Supernova Cosmology Project and High-Z teams claim they are 99 percent confident that the brightness discrepancy is real, but they grow cagier when asked if the discrepancy proves we live in an accelerating universe. Their findings rely on a tremendous leap of faith in the knowability of the universe and in the reliability of their understanding of it. James Peebles at Princeton, who has watched lots of cosmological fads come and go over the years, said, "I would give better than fifty-fifty odds of Lambda having been detected. But I wouldn't give better than three-to-one."

For one thing, nobody really understands how Type Ia supernovas work. When astronomers look at distant supernovas, they are seeing them as they were billions of years ago when they lived and died; it has taken all this time for their light to reach telescopes on the earth. Did those ancient stars behave the same way as the much better-studied nearby ones? Might the stars themselves have been fainter in the past? Perlmutter's and Schmidt's teams have buried themselves in supernova research to make sure Mother Nature wasn't fooling them. But supernovas are incredibly powerful, intricate things. The closer scientists look, the more mysterious they seem.

The detailed saga of a Type Ia supernova goes something like this. A modest, sunlike star in a double-star system grows old. It puffs up to become a swollen red giant star and then blows off its distended outer parts. The hot, extremely dense cinder left behind is the white dwarf. It is composed mostly of carbon, nitrogen, and oxygen, nuclear by-products from the fusion reactions that powered the star. Because its mass is so tightly packed, the dwarf generates an intense gravitational field that can strip gas from the other star. It continues to gobble greedily until it porks up to a mass of 1.4 times that of the sun. At that point, the dwarf's gravity abruptly overwhelms the electrical forces between electrons and protons, which previously kept it from collapsing further. In an instant, the star implodes and grows furiously hot. A wave of nuclear fire tears through the star and unleashes so much energy that the star blows to bits, flying apart at more than five thousand miles per second. This pyrotechnic cloud of radioactive debris produces the brilliant light of the supernova.

Type Ia supernovas all start as the same basic kind of star—a white dwarf of exactly 1.4 solar masses—so intuition tells you that the explosions should all be more or less the same. But intuition is a shaky basis for publishing in the *Astrophysical Journal,* and the current theoretical models of supernovas have more holes than a prairie dog town. First off, the details of what goes in that nuclear flame are far too complicated to simulate. Kirshner launches into a sardonic David Letterman–style monologue as he pokes fun at how little astrophysicists know about these details. "It happens in a flame whose thickness is millimeters, and the amount of burning depends on the topology of the flame. Give me a break. Can they calculate this? Nooooo!" he says, laughing. Furthermore, there is a woeful lack of information to feed into the theories. No one has ever observed the star that becomes a Type Ia supernova before it pops off, so all the ideas about them are inferences. It is like trying to figure out how to build an A-bomb from watching footage of a nuclear test.

Even if the laws of physics haven't changed, the stars themselves have. Billions of years ago, the chemical makeup of the universe was different from what it is today. Galaxies had not been so thoroughly polluted by the heavy elements that spew from giant stars, novas, and supernovas. The stars within those distant galaxies were also different, presumably. So Perlmutter and Schmidt have studied supernovas in many astronomical settings—in youthful spiral galaxies, where new stars still are forming, and in mature elliptical galaxies, for instance. On the whole, the explosions all seem the same. But Adam Riess at Berkeley, working with Schmidt, spotted a possible sign of trouble. He found that distant explosions reach their peak noticeably faster than nearby ones. Such hastiness doesn't necessarily mean the distant ones were fainter, but it does suggest that astronomers cannot safely assume that Type Ia supernovas all live by the same rules.

The other worrisome source of error is almost insanely simple: dust. Maybe the only reason the distant supernovas look fainter is that their light is absorbed by intervening, opaque material. Perlmutter and Schmidt have done everything they could to correct for this effect. Small dust particles scatter blue light more than red, so dust should redden the apparent colors of the supernovas. Despite careful studies, neither team sees any sign of such color changes. Of course, nature could be perverse and have filled the universe with a different kind of dust that scatters all colors equally. Such dust would be very hard to spot. If you had been wearing a pair of sunglasses all your life, you might never realize that you weren't seeing the world's true appearance.

One way to address these concerns is to look at extremely distant supernovas. Up to a distance of about six billion light-years, Lambda makes supernovas fainter, because space has been expanding more quickly than expected. At greater distances, the effect changes direction. We're now seeing back to a period in the universe when Lambda has not had as much time to take effect and the universe is expanding more slowly than expected. In other words, the most remote, highly redshifted supernovas should be brighter than they would be without Lambda mucking up the works. If dust is blocking the light, or supernovas were fainter long ago than they are today, the effect should keep getting worse as astronomers look over greater distances, further into the past.

In October 1998, Perlmutter's team found one such distant supernova, nicknamed "Albinoni," using the Keck II telescope, one of the few large enough to peer out to such distances. The name comes from Perlmutter's fondness for classical music and aversion to the usual impersonal number-and-letter designations that have come to dominate astronomy. Albinoni lies nearly ten billion light-years from the earth and appears a little on the bright side, just as Perlmutter hoped it would. At the January 1999 American Astronomical Society meeting, on the first anniversary of the momentous announcement, Perlmutter addressed a packed auditorium and sported a hopeful grin when he pointed to the dot on his graph indicating Albinoni. He saw it as one more sign that the big bang is speeding up. In April 2001, Riess analyzed an even more distant Type Ia captured by the Hubble Space Telescope. This star, too, appeared brighter rather than dimmer. Even doubting Thomases began to believe in dark energy.

In fact, most astronomers embraced the new gospel almost immediately. "It's remarkable how little intelligent criticism there's been," says Kirshner. He notes that the most intensely hostile comments have come from within the two supernova groups. For a while he told his fellow team members, "In your heart you know it's wrong." Yet the outside reception could hardly have been warmer. The current generation of astronomers has grown up with competing cosmologies and with constant reminders that attaining cosmic enlightenment means recognizing there is more to the universe than meets the eye. Perlmutter and Schmidt were confused, but even more they were elated. The whole reason for embarking on such a grueling endeavor was to find something inexplicable.

Even more striking was the reaction of the theoretical community, the deep thinkers of sci/religion who follow in Einstein's tradition of building the universe with the brain, an old envelope, and a good ballpoint pen. The supernova results brought them some incredibly good tidings. For years the observers had been telling the theorists that the density of the universe is much smaller than required by inflationary theory, which firmly predicts that the shape of space is flat and hence the density must lie exactly at the critical point between permanent expansion and eventual collapse. Surveys of the distribution and dynamics of clusters of galaxies put the true matter density much lower, at something more like 0.2 or 0.3 of that critical amount. Then along came Perlmutter and Schmidt, whose findings say there is exactly enough vacuum energy to place the total cosmic density where the inflationary cosmologists predicted (and hoped) it would be. "I thought those guys would take great pleasure in this, say, 'We have been right all along, this is the proof.' But it's not true," Kirshner says with a bemused laugh. Guth listened to back-to-back presentations by Kirshner and Perlmutter at the 1999 American Astronomical Society meeting and reacted with a shrug. "It doesn't change things much for inflation," he said nonchalantly.

Einstein's followers adored the beauty of an inflationary universe whose density is very close to the critical value. They felt inflation had to be true, just as the great prophet felt light had to bend around the sun. "If you have a good theory, you pursue it until the data rule it out—and inflation was a good one," says Turner. By the early 1990s, a number of theoretical cosmologists had decided that they needed Lambda back in their equations. It was the only thing that made their otherwise beautiful readings of cosmic scripture make sense. "Everyone was holding their noses, because of the checkered history of Lambda," says Turner. "There were a few rats that jumped ship, but most of the rats stayed on board." When Perlmutter and Schmidt spread the word that they had found signs of Lambda, Turner was pleased but not exactly astounded. For nearly ten years the theorists had known that their models worked best when cooked up with a dash of Lambda.

But this was more than blind faith. Lambda had changed a great deal since 1917. When Einstein conceived of it, Lambda was purely a philosophical invention. Ever since Yakov Zeldovich explored the connections between cosmology and quantum physics, however, it had become an increasingly testable piece of the overall sci/religious description of the world. The virtual particles that constantly pop in and out of empty space, bizarre though they may seem, have measurable effects. Since the mid-1990s there's been another tangible and very persuasive argument in favor of Lambda. When Guth and Linde originally assumed that the universe had exactly the critical density, there were no observations to back them up. Now there are.

Once again, the most decisive evidence in cosmology comes from studies of the cosmic microwave background, the echo of the big bang. Cosmologists have reconstructed the early history of the universe with great precision and concluded that this background dates from the time when the universe was three hundred thousand years old, when matter cooled enough to form atoms and suddenly turned transparent to radiation. It is fairly easy to calculate the size distribution of the structures that could have formed in that time. The overall mix of large and small lumps in the early universe, revealed in 1992 by the COBE satellite, accurately matches the pattern predicted by inflation. Moreover, the largest structures serve as giant measuring sticks in the sky. If the geometry of the universe is flat, they should show up in a microwave map of the sky as markings about one degree wide, or about twice the width of the full moon. If space is curved, the markings should look distinctly smaller or larger, depending on the type of curvature.

Studies from the two balloon-borne microwave telescopes—BOOMERANG and MAXIMA—show a pattern consistent with a flat, critical-density universe. That finding offers more support for inflation "There are no noticeable discrepancies," Guth says. More recently, the two balloon experiments have found secondary fluctuations in the microwave background that look like the effects of hot matter flowing into and sloshing out of dense regions in the early universe. These fluctuations imply a balance between matter and dark energy that exactly matches the supernova results.

As a result of these developments, Lambda is not only acceptable, it is downright fashionable. Cosmologists no longer need to whisper that they are seeking the secrets of the Old One. Yet the quest toward cosmic truth is far from over. Einstein didn't know what Lambda was when he stuck it in his equations, and we still don't know. Cosmologists talk in vague terms about how Lambda might result from the vacuum energy predicted by quantum physics. But as physicists have noted for a quarter century now, quantum theory calls for a vacuum energy far, far beyond what would be contained within Lambda. So much energy would literally blow the universe to bits. That leaves cosmologists in the awkward position of arguing that there is some mechanism, entirely mysterious, that cancels out all but one part in 10^{120} of the quantum vacuum energy, leaving just enough to make the current picture of the universe work properly. That is why Guth initially found the supernova results shocking—not because they indicate Lambda exists, but because Lambda is so small compared to the proto-Lambda that powered inflation. "We've got to learn more about what this dark energy is. There's nothing more fundamental than figuring out the energy that dominates the universe," Perlmutter says.

It seems doubly odd that Lambda is roughly equal in density to the matter in the universe—within a factor of two, at least. Matter gets diluted as space expands, but each bit of new space contains more Lambda, which begets more Lambda, and so on. Early in the universe, the matter density would have been millions of times as great as Lambda. In the far future, the roles will reverse. Right now, for a very short period in the overall history of the universe, the two are roughly equal, so the cosmos is experiencing only a subtle acceleration. Schmidt finds the coincidence hard to swallow: "It's just bizarre to me that by chance we're living in a time when these things are roughly balanced. If we'd lived five billion years ago, we couldn't even measure Lambda, and five billion years in the future it would be screamingly obvious." This is exactly the kind of fine-tuning inflation was supposed to get rid of.

Always eager to press on toward deeper cosmic understandings, theorists have cooked up a new kind of dark energy that has more appealing attributes than the vacuum fluctuations, called "quintessence." The name refers back to the Greek theory of matter, in which the earth is composed of four basic elements but the heavens of an elevated, more perfect "fifth element." The name deliberately evokes Aristotle's ether, describing that extra intangible that explains why the universe moves. In its modern incarnation, quintessence is something that is not matter but isn't exactly energy in the usual sense, either. Paul Steinhardt at the University of Pennsylvania, who collaborated with Guth in creating a workable theory of inflationary cosmology, is the leading champion of quintessence. Even he has a hard time explaining what it is in simple terms, however: "a slowly varying, spatially inhomogeneous component" or "a scalar field slowly evolving down its potential." Again, the words are new, but the impulse is old. It is the endearingly optimistic sci/religious hope that one more mystical element will pull the whole cosmic explanation together.

A scalar field is one that has magnitude but, unlike an electric field, for instance, has no direction. The energy just sits there, piled up in space. It is somewhat analogous to the electric charge that builds up when you walk along a carpet on a dry winter day. Your body is full of energy, but you don't feel it. It just stays in place, at least until you disturb the system by doing something like touching a metal doorknob and get a nasty shock. According to the latest thinking in physics, space may be full of energy in the form of scalar fields, which regulate the rules governing subatomic particles and may even give those particles their masses. Indeed, the energy that presumably caused inflation in the early universe came from a scalar field.

The primary appeal of quintessence is that it does not have to stay at a fixed value, unlike vacuum energy. In some of Steinhardt's models, quintessence decreases in energy as the universe expands, tracking the thining out of the universe. It may not be a coincidence that the cosmic matter and energy densities are similar; they might be linked to one another physically. And Steinhardt does away with the mystery of how the vacuum energy almost, but not exactly, shrinks to zero. He simply argues that the energy really is zero because of some fundamental property that cancels out the effects of the quantum fields, as most physicists have long assumed, and proposes instead that quintessence drives the observed cosmic acceleration.

Even by today's cosmological standards, this is deep prayer in the Church of Einstein: unknown effects controlling the vacuum energy, unknown fields creating quintessence, unknown effects linking quintessence to the matter density of the universe. Hardcore observers like Kirshner bristle when Steinhardt's name comes up. "Some of us would like some scintilla of evidence," he crows.

Yet the two men share a similar core of epistemological faith—Steinhardt that he can deduce ultimate truth, Kirshner that he can hope to see it.

Further studies of supernovas from the ground will not help much here. Cosmologists are heavily pinning their hopes on two new orbiting telescopes. NASA's Microwave Anisotropy Probe, launched in 2000, is already putting cosmology to the test as never before. The European Space Agency's *Planck* spacecraft will follow up in about five years. These satellite experiments contain instruments to scrutinize the cosmic microwave background in excruciating detail. The resulting maps should be able to distinguish between quintessence and Einstein's Lambda and to throw out some of the inflationary cosmology models by revealing a great deal about the state of the universe very soon after the big bang. Meanwhile, the various supernova researchers are promoting SNAP—the SuperNova/Acceleration Probe. If approved, this satellite will provide vastly improved measurements of Lambda and other astrophysical phenomena. The new data will roll back barriers, increase the authority of sci/religion, and, no doubt, spawn even more ambitious and elusive theories.

For now, Perlmutter and Schmidt can say only what the universe appears to be doing, not why it is doing it. But taken at face value, the "what" alone is plenty strange. They have found a kind of cosmology that even the most far-out theorists had not seriously considered before. When Alexander Friedmann codified the first mathematical descriptions of a dynamic universe, he established the mathematical foundation for three basic descriptions of how it might evolve over time: the universe might collapse again someday, it might steadily expand forever, or it might teeter just on the edge between the two. Until 1998, those were the only scenarios that most cosmologists took seriously; in *A Brief History of Time*, updated in 1996, Hawking presents these essentially as the only possible choices. When Einstein invented the cosmological constant, however, he unwittingly established a fourth spiritual possibility. He had imagined that Lambda would exactly balance the force of gravity, keeping a finite universe motionless and intact. What Perlmutter and Schmidt see, instead, is the ultimate in unbalance—a runaway universe, in which galaxies race ever faster away from one another. The big bang gave sci/religion its own creation story. The discovery of Lambda offers a possible look at the end of days.

If cosmological acceleration continues unchecked, it holds out no hope of rejuvenation and new beginnings. Rather, it leads to an increasingly empty and isolated future, where stray clumps of matter are surrounded by furiously expanding space, racing away to nowhere. "It's going out in the bleakest fashion I can think of—it's eternity, but it's nothingness at the same time. If the universe had too much matter in it and recollapsed into a 'gnab gib' [backward big bang], that's at least exciting and has a finite end—it's like death. This other thing is just really grim," Schmidt says. The accelerating universe could even create a spiritual crisis for Einstein's brand of sci/religion. He had originally posited a finite universe because it followed Mach's view that inertia is caused by the way an object interacts with the combined gravitational field of the rest of the universe. So what happens when the rest of the universe is no longer visible? "If you believe in Mach's principle, you can wonder if inertia will start not to make sense when the expansion takes over. It might take another Einsteinian revolution to answer that," Perlmutter says.

From his viewpoint, Perlmutter isn't so worried about the prospect of a speeding-up cosmos. It allows an endless amount of time for things to happen, he notes. Perhaps we just need some time to adjust our philosophical attitudes to the latest scientific news. For old-time religions, however, he sees the accelerating universe deflating the stale argument that modern cosmology is just recapitulating the biblical story of creation: "The accelerating universe loses that sense that everybody likes to have that once you find out, science will turn out to be just the same as somebody's myth." Riess captures the mix of wonder and accessibility that is the hallmark of sci/religion: "When I tell people about our different possible fates, it sounds philosophical—like talking about God. But this is something you can measure, like weather forecasting."

CHAPTER NINE

SALVATION IN THE CHURCH OF EINSTEIN

"YOU WILL HARDLY find one among the profounder sort of scientific minds without a religious feeling of his own," Einstein wrote in 1934. "This feeling is the guiding principle of his life and work . . . It is beyond question closely akin to that which has possessed the religious geniuses of all ages." A dozen years ago, I set out in search of that modern spiritualism at meetings of the American Astronomical Society. Every six months approximately 2,000 working astronomers—the country's largest gathering of cosmic explorers—take over a convoluted network of conference rooms and spend five days updating each other on the state of the universe. The look of these convocations, I soon discovered, is consistently and unnervingly businesslike: worn red carpeting, windowless rooms with yellowish overhead lighting, and participants scurrying around with plastic-encased nametags clipped to their jackets. More disappointing, the prevailing tone of the presentations is technical and detached. Many of the most renowned priests of sci/religion show up, but they offer little insight into the feeling that motivates their research. And I have never, ever heard them utter the word "God."

Alan Guth, whose elaboration of the big bang looms large over the Astronomical Society meetings, regards this reserved style as a sign of the times. "There's been a change in how physicists talk. Einstein felt free to talk about God; physicists today abhor the word," he says. The shift is paradoxical: As the Church of Einstein has grown steadily more powerful and compelling, its congregation has renounced much of the vibrant language it borrowed from old-time religion. Einstein's cosmic religion, his sense that the workings of the universe are harmonious and ultimately comprehensible to reason, is now buried within terms such as string theory or supersymmetry. Gamow's *ylem* has become quark-gluon plasma. Lemaître's brightly evocative "fireworks of creation" has become the arcane "cosmic microwave background, left over from the time of recombination." Even the big bang, whose name vaguely connotes the story of Genesis, is commonly deconstructed into more precise but puzzlingly abstract terms—a primordial quantum fluctuation, an inflationary episode, parity violation.

Obscurity of language is now practically a defining trait of modern cosmology, much as it is in particle physics, the branch of science operating at the other extreme end of the scale. Such obfuscation can make the gospel of sci/religion seem as remote as Christian heaven or Aristotelian ether. Small wonder, then, that much of the public has only a fuzzy understanding of the mystical program of sci/religion. A community of science reporters, myself included, makes a living decoding the esoteric utterances of leading cosmologists.

Prayer in the Church of Einstein seems a daunting task, possible only after mastering a litany of unfamiliar terms and the enigmatic mother tongue of mathematics. Wander from room to room at the Astronomical Society meeting and you might never even know the faith is there. At a January 2002 session, a young astronomer named Brian Mason, of Caltech, presented the lead paper in one of the main cosmology sessions. His discussion about ways to measure the physical conditions in the early universe bore the title,

"Measurements of the CMB Power Spectrum to L = 4000 with the CBI." Compare that to the blunt directness of Einstein's "Cosmological Considerations on the General Theory of Relativity" or Gamow's 1950 lofty paper summarizing how elements formed during the big bang, "Half an Hour of Creation." The language of science sounds as alien as church Latin or biblical Hebrew.

As the sermons of sci/religion have grown more oblique, its temples have similarly become more remote and less romantic. When I drove to the top of Mount Wilson, the site of cosmology's greatest early observational triumphs, I encountered a set of buildings that look decidedly past their prime. The visitor's center smells musty and needs a good dusting. Hubble's bentwood chair sits forlornly beneath the antique-looking Hooker telescope. It's not even his real chair, but a knockoff placed there in his memory. The cutting-edge observatories have retreated to higher locations, scraping the stratosphere on Mauna Kea in Hawaii, Cerro Paranal in northern Chile, and a handful of similarly inaccessible peaks where visits are expensive, time-consuming, and physically taxing. The Hubble Space telescope, probably the most famous sci/religious tool in the world, can be reached only by a $400 million trip on the space shuttle. Astronomers are no longer the solitary dreamers spending long, shivering nights peering through an eyepiece. Silicon light detectors and high-speed phone lines carry the images to computer labs that are halfway down the mountain or halfway across the world, where the real observing takes place. Ambitious projects such as Perlmutter's supernova search go further, relying on electronic processors to make the initial interpretation of whether there is anything interesting in each image, because the human eye could not possibly keep up with the flow of data.

These twin developments have created the gross misimpression that sci/religion is far removed from the lives and concerns of the lay public. Cosmologists do not ascend to flee the flaws of the world we inhabit, like saints striving to soar toward heaven. Rather, they ascend in order to bring the cosmic truth down to Earth and erase the boundaries between the universe up there and the world down here, boundaries that have existed only in our minds and in our religious tales. And the unfamiliar vocabulary of sci/religion is not intended to create an insular, dogmatic faith, the way that the Bible or Koran and their elaborate body of interpretive literature define the practice of Christianity and Islam. Of course there is an element of clubbiness to the way cosmologists speak. They have their own shop talk, just as lawyers and auto mechanics do. But their jargon ultimately serves the same goal as their observatories, to unify us with our universe. The terminology of sci/religion provides a detailed way to describe wonders that are beyond normal human experience, both in space and in time. These words describe Einstein's divine vision of a coherent universe ruled by general relativity and other knowable physical laws.

While I toured Mount Wilson, I had to keep readjusting my opinion of what I was seeing, shedding the distorting attitudes of the ancient beliefs. For old-time religion, the decline of such a sacred site would be cause for mourning. For sci/religion, however, the decline of Mount Wilson is actually cause to rejoice. The Church of Einstein has no holy ground; it operates on faith alone. Its temples of observation, like the theories they bolster or refute, are therefore in a continual state of flux. Old instruments are downgraded or discarded as new ones come along, taking the followers to a finer resolution, a better sensitivity, or a previously unexplored region of the spectrum. Mount Wilson grew quiet as larger telescopes utilizing superior technologies opened on darker, higher, more isolated peaks. But that is an old and ongoing story. The sites of all of the greatest sci/religious discoveries witness continuous change. Mauna Kea now has twelve working telescopes at its frigid summit. Five of these, including the magnificent Keck twins, began operation just within the past decade. Photon by photon, these observatories are continuing the job of the Hooker telescope, not to mention William Herschel's wooden-tubed reflector and Galileo's spyglass. They are bringing us closer to a transcendent connection with the cosmos.

Moreover, the temples of sci/religion can experience a renewal of purpose that has no analogy among the old-style faiths. The Hooker telescope is undergoing a modest rebirth with the addition of adaptive optics, a system of bendable mirrors and computer electronics that cancels out atmospheric blurring. And all across the top of the mountain, shiny new pipes link together the CHARA array—a network of modest, one-meter (three-foot) telescopes that combine their light to produce the resolving power of a single mirror as wide as the space between them. This technique, called "optical interferometry," can produce images hundreds of times sharper than those from the Hubble Space Telescope. It only works on bright sources, so interferometry won't investigate Lambda or monitor the evolution of distant galaxies. But it can watch stars and planetary systems forming, Cepheid variable stars pulsating, perhaps even infernally hot gas taking a one-way trip into a black hole. Although such discoveries do not expand the boundaries of the universe, they do make more of its invisible wonders manifest.

At the American Astronomical Society, as well, there is more sci/religious zeal than the superficial sights and sounds indicate. Despite their reluctance to invoke overtly theological language, in their actions the current priests of sci/religion remain firmly committed to the Church of Einstein. "I don't think the basic attitudes have changed. Modern physicists, like Einstein, have their own intuitions about how the world works," Guth says. Mason's paper at the society's recent meeting is a strong case in point. The "CBI" in his paper is the Cosmic Background Imager, an ingenious array of detectors designed to pick out features within the glow of microwaves left over from the big bang. Patterns embedded in the glow indicate how matter and energy were distributed when the universe was just 300,000 years old; that information, in turn, can reveal the constitution of the universe and distinguish between different models of early cosmic evolution. In other words, Mason is continuing to expand the kingdom of Einsteindom by pressing our physical understanding farther out into space, further back into time.

During an afternoon lull at the Astronomical Society meeting, Perlmutter laughed about how quickly his colleagues have absorbed the accelerating universe into their cosmological theories and transformed his once-shocking discovery into an element of the conventional wisdom. So he is pressing on ahead, gathering more data, planning additional observations, and delegating some of the computational problems in order that he might carve out some time to sleep and—if he is lucky—dig through some of his enormous backlog of e-mail. He's also working up support for building that orbiting observatory, the SuperNova/Acceleration Probe (or just SNAP), which would watch for exploding stars from space and drastically improve on the measurements possible from the ground. There's a lot of bureaucracy involved in obtaining the tens of millions of dollars needed for such a mission. "It's like review-of-the-month club right now," he says. Still, he remains boyishly giddy about the need to understand the intangible. "We've got to know more about what this dark energy is—nothing is more fundamental than figuring out the energy that dominates the universe," he says. He never mentions God, but his words thrum with the heart-pounding promise of Einstein's cosmic religion.

Thus at the top of Mount Wilson or in the basement conference rooms of the Washington Hilton, the challenge to sci/religion is the same. The Church of Einstein is more authoritative and comprehensive than ever. Its spiritual power has eclipsed that of the old-time religions. And yet by and large the public does not appreciate the full mystical dimension of modern science. They certainly understand and admire its practical results, ubiquitous in technological advances such as cable television, GPS-guided maps in cars, and MRI scans for surgery. They generally appreciate the breadth of its explanations as well. These days news about planets around other stars or the mass of the neutrino finds more outlets than ever in the papers, on television, and on the Web. Nevertheless, there is a widespread perception that science is limited. The familiar line is that science has nothing to say about morals or about the purpose of life—in short, that it is not enough like old-time religion. Sci/religion's unfamiliar terminology and specialized tools of research contribute to the view that it is aloof to these human hungers. Interestingly, the scientists themselves often knowingly help foster that impression.

Many researchers applaud the limited definition of the scope of science because it helps maintain the delicate cease-fire that Galileo attempted to strike nearly four hundred years ago when he cited Cardinal Baronius's dictum that the Bible reveals "how one goes to heaven, not how the heavens go." Dividing the world in that manner assures that the Pope will never again attempt to dictate the mechanical workings of the universe. In the years since Einstein's first cosmology paper, the Catholic Church has made peace with scientific explanations of the origin of the universe and with most of evolutionary biology; it has even apologized for censoring Galileo. All that scientists have had to give up in return is a claim to authority on salvation, the afterlife, moral behavior, and other topics that they never presumed to talk about in the first place. This modern line of demarcation also serves as a warning to creationists: Don't interfere with our descriptions of the material world, and we won't interfere with your views on purely theological matters. The one huge problem is that this solution requires abnegating the spiritual dimension of science. It denies the very existence of sci/religion.

Robert Kirshner is an outspoken proponent of this two-worlds interpretation. "I don't see what cosmology has to do with religion, except that it has to do with the beginning of things and the end of things. It certainly doesn't tell you anything about what most people do as religious practice, which has more to do with how to be as a person," he says. Kirshner considers the trade-off a good bargain: "Since Galileo, the Church has gotten smarter and realized they shouldn't hang religious belief on some particular question about the physical nature of the world. So now it doesn't matter what we do." The late Stephen Jay Gould of Harvard, a prominent evolutionary biologist and prolific science popularizer, largely echoed this view. "Science tries to document the factual character of the natural world, and to develop theories that coordinate and explain these facts. Religion, on the other hand, operates in the equally important, but utterly different, realm of human purposes, meanings, and values—subjects that the factual domain of science might illuminate but can never resolve . . . I propose that we encapsulate this central principle of respectful noninterference . . ." he wrote in his most recent manifesto on the two worldviews.

Other scientists, especially those on the theoretical edge of cosmology, refuse to abide by that noninterference pact but nonetheless have an acute sense of a boundary between science and traditional religion. They see the expanding domain of sci/religion as theologically significant because it keeps eliminating places where God could hide, or ways in which a willful deity could exert fundamental control over the operation of the universe. In a sweeping sci/religious meditation entitled "Universe, Life, Consciousness," Andrei Linde states: "The possibility that the universe eternally re-creates itself in all its possible forms does not necessarily resolve the problem of creation, but pushes it back to indefinite past. By doing so, the properties of our world become totally disentangled from the properties of the universe at the time when it was born (if there was such time at all). In other words, one may argue that the properties of our world do not represent the original design and cannot carry any message from the Creator." Pressed on his own beliefs, Linde equivocates. "It doesn't mean that there is no place for God, just that there are some new possibilities. I am not a religious person or an antireligious person; I am just trying to figure out what I can say without emotion," he says.

Stephen Hawking is by far today's most famous proponent of this program to smoke God out of His potential hiding places. Hawking's no-boundary proposal—a ferociously complex argument that there was no first moment in time, even though the universe has a finite age—is explicitly designed to eliminate the usual need for a first cause. "[I]f the universe is really completely self-contained, having no boundary or edge, it would have neither beginning nor end: it would simply be. What place then for a creator?" he asked in *A Brief History of Time.* At the end of the book he attempts to answer his own question. He envisions that scientists might figure out a physics theory that brings about its own existence. "Then we shall all, philosophers, scientists, and just ordinary people, be able to take part in the discussion of the question of why it is that we and the universe exist. If we find the answer to that, it would be the ultimate triumph of human reason—for then we would know the mind of God." Broadly speaking, he expresses the same sentiment that Einstein voiced when he yearned to know the secrets of the Old One. But Hawking shies away from invoking anything like the "cosmic religious feeling" that Einstein saw as an essential guide to scientific inquiry.

Hawking-style efforts to reduce the authority of the old-time Gods don't attract a lot of converts because of this failure to offer a new spiritual program. Perlmutter sees science and religion continuing to run on parallel tracks. "As you go farther back in time and you study more about how the world started, it doesn't displace God because you still have to say, well, what began all these things? Why is there something rather than nothing? Why are there these laws rather than no laws? I think for the people to whom God gives an answer for that, it still gives an answer, and for those to whom God doesn't seem like an answer, they still have to ask, where does God come from?" he says. Conversely, he paints a playful picture of how scientists could remain detached in the face of divine revelation: "Even if you were able to show scientists the existence of God, it still doesn't stop the scientific process. Just as knowing more about the world doesn't stop the possibility of there being a God, it doesn't make a difference the other way around either, because you still want to do the science. If you found out that miracles really do happen and God's creating them, probably the immediate response of science would be, well, how does God do that? Is He doing it within the laws of physics, or does he get to change the laws of physics momentarily?"

When scientists are willing to argue on behalf of Einstein's old sense of cosmic religion, they often do so gently, almost apologetically. Asked about his personal faith, Brian Schmidt responds, "My grandparents were very religious and I was always exposed to religion, but it's something that never really took with me. The problem is, to have religion you have to have faith in certain principles. But the principles I adopted were faith in physical law, not faith in a supreme being. My attitudes toward cosmology has been formed by this faith in physical law—which is somewhat religious at some level, but it is not what we think of conventionally as religion."

Among the many voices, a handful of the sci/religious faithful have spoken out forcefully about the mystical glory of their work. Carl Sagan greatly advanced the cause through his popular books and his television series, *Cosmos.* His oft-repeated epigram, "we are made of starstuff," neatly summarizes the sense of cosmic unity that springs from the theories pioneered by Einstein, Lemaître, and Gamow. From outside the world of cosmology, the British evolutionary biologist Richard Dawkins sermonizes on the ability of sci/religion to stir the soul. In 1996 he delivered an electrifying presentation to the American Humanist Association, crisply titled "Is Science a Religion?" His short answer was no, because its methods are so different. His long answer, however, addressed the ability of science to fulfill the human hunger for explanation that old-time religions have long sated: "All the great religions have a place for awe, for ecstatic transport at the wonder and beauty of creation. And it's exactly this feeling of spine-shivering, breath-catching awe—almost worship—this flooding of the chest with ecstatic wonder, that modern science can provide. And it does so beyond the wildest dreams of saints and mystics. . . . The merest glance through a microscope at the brain of an ant or through a telescope at a long-ago galaxy of a billion worlds is enough to render poky and parochial the very psalms of praise."

Joel Primack, a cosmologist at the University of California, Santa Cruz, who studies the formation and evolution of galaxies, offers an unusually trenchant description of the relationship between sci/religion and its wobbly predecessors. "Rather than assuming science and spirit are separate jurisdictions, I assume that reality is one, and that truth grows and evolves with the universe of which it speaks," he writes. "Every religion is a metaphor system, and like scientific theories, every religious myth is limited. Perhaps progress in religion can occur as it does in science: without invalidating a theory, a greater myth may encompass it respectfully, the way general relativity encompasses Newtonian mechanics. In the next few decades, powerful ideas of modern cosmology could inspire a spiritual renaissance, but they could also be totally ignored by almost everyone as irrelevant and elitist." If Einstein's cosmic religion fails to find an audience, that will be a terrible loss not just for the devout researchers on a quest for ultimate knowledge, but for everyone seeking a more just and peaceful world.

More than three hundred years ago, the rationalist philosopher and theologian Baruch Spinoza described the dangers inherent in the biblical description of an interventionist God who responds to prayer. This description led people to believe that God acts to bring about certain goals, a notion Spinoza considered inherently corrosive. "Everyone thought out for himself, according to his abilities, a different way of worshipping God, so that God might love him more than his fellows, and direct the whole course of nature for the satisfaction of his blind cupidity and insatiable avarice. Thus the prejudice developed into superstition, and took a deep root in the human mind," he wrote. Because God did not reliably respond to these prayers, worshippers adjusted their superstition and came to believe that the Lord's judgments are beyond human understanding. "Such a doctrine might well have sufficed to conceal the truth from the human race for all eternity, if mathematics had not furnished another standard of verity in considering solely the essence and properties of figures without regard to their final causes," Spinoza concluded.

Einstein repeatedly aligned himself with Spinoza's philosophy and concerns, identifying the "religion of fear" as a primitive and destructive stage in the spiritual development of humankind. "I cannot imagine a God who rewards and punishes the objects of his creation, whose purposes are modeled after our own—a God, in short, who is but a reflection of human frailty," Einstein wrote in 1932. Nine years later, in a paper contributed to a symposium on science, philosophy, and religion, he expounded on the stunting effects of belief in an all-knowing, personal deity. Such a God conflicts with the demands of science by potentially interfering with the predictable operation of natural law. Such a God also conflicts with the needs of humanity. "If this being is omnipotent, then every human action, every human thought, and every human feeling and aspiration is also His work; how is it possible to think of holding men responsible for their deeds and thoughts before such an almighty Being?" Einstein asked. He concluded that "[t]he further the spiritual evolution of mankind advances, the more certain it seems to me that the path to genuine religiosity does not lie through the fear of life, and the fear of death, and blind faith, but through striving after rational knowledge."

Dawkins, probably the most outspoken critic of old-time religion among the current generation of scientists, pushes this line of attack much farther. Again, from his classic 1996 speech, words that still have the old-time faithful sputtering: "It is fashionable to wax apocalyptic about the threat to humanity posed by the AIDS virus, 'mad cow' disease, and many others, but I think a case can be made that faith is one of the world's great evils, comparable to the smallpox virus but harder to eradicate. Faith, being belief that isn't based on evidence, is the principal vice of any religion. And who, looking at Northern Ireland or the Middle East, can be confident that the brain virus of faith isn't exceedingly dangerous?"

Fortunately, sci/religion doesn't need to make its case solely by tearing down the ancient faiths. Sci/religion offers a positive and immensely appealing alternative way to look at the world, a religion of rational hope. It covers much more than "how the heavens go," although that aspect is the foundation on which all its other elements are built. Our current picture of the big bang—fluctuating out of nothingness, filling with particles and radiation, inflating swiftly and then evolving over billions of years, under the guidance of unseen dark matter and dark energy, into its present form—is the most complicated, comprehensive, and thoroughly tested creation story in history. And like all such stories, the big bang tells a lot about who we are and how we think of ourselves. We are no longer content with the revealed truths of Genesis and other ancient mythologies. We want to participate in the origin of the universe by understanding it on our own terms and connecting it to laws and phenomena that we can study. Old-time religion delivers only a single, unchanging version of the creation story.

Sci/religion recognizes our human limitations and provides an opportunity to keep challenging, discarding, refining, and updating our cosmological models in the endless search for truth.

Just as sci/religion's picture of creation surpasses that of the ancient mythologies, so its process of revelation takes prayer to a new level. Einstein retraced the scientist's discovery process in poetic, near-ecstatic terms. "His religious feeling takes the form of a rapturous amazement at the harmony of natural law, which reveals an intelligence of such superiority that, compared with it, all the systematic thinking and acting of human beings is an utterly insignificant reflection. This feeling is the guiding principle in his life and work, in so far as he succeeds in keeping himself from the shackles of selfish desire. It is beyond question closely akin to that which has possessed the religious geniuses of all ages," he wrote in his 1934 book *Mein Weltbild.*

And still that is only part of the redemptive power of sci/religion. Notwithstanding the claims of old-time religion and some logic-toting atheists, the Church of Einstein does have something to say about how to go to heaven—not as a guide to the afterlife, but metaphorically as a guide to living a moral life. The value system shows up most clearly in cosmology and in sci/religion's other book of the creation story, evolutionary biology. Both disciplines emphasize the unity of our species, with the universe and with each other. Evolution demonstrates that all humans are closely related: *Homo sapiens* has existed for fewer than 175,000 years and is extraordinarily uniform at the genetic level. Cosmology shows that we are all interconnected at a more glorious and humbling level as well. We share a tiny, fragile, and precious refuge, the only habitable planet that we know of. All of us are made of the same elements, forged in the same stars, governed by the same miraculous physical laws that allow the sun to shine, rivers of water to flow, and atoms of carbon to bond with hydrogen and oxygen in our chemically dynamic bodies.

Sci/religion is a human faith, prone to distortions and misinterpretations, but it naturally inclines toward tolerance and environmental sustainability. It is inherently democratic, holding that cosmic truth is available to all through open inquiry, not just to the few who adhere to one particular theology or ideology. No wonder scientific progress stalled in Nazi Germany and Soviet Russia. And the sci/religious exploration of the universe unequivocally requires abandoning the idea of a personal God who can be summoned and commanded by prayer. That abandonment again bolsters the cause of unity by undermining the traditional, religious motivations for aggression and hatred.

Sci/religion invokes another kind of belief in its place. "It is the aim of science to establish general rules which determine the reciprocal connection of objects and events in time and space. For these rules, or laws of nature, absolutely general validity is required—not proven. It is mainly a program, and faith in the possibility of its accomplishments in principle is only founded on partial successes," Einstein wrote. In contrast to the old-time religions, sci/religion has mounds of evidence showing that its assumptions are correct. But the doctrine of falsification through observation forces sci/religion to be honest and admit that it can never state with total certainty that the same physical laws apply in all places at all times. There could always be an unseen exception, or simply one that has not occurred yet.

There is another way in which sci/religion helps confront the problem of human aggression. The new faith is rooted in curiosity and examination. It can grab control of the old human conquering spirit—a relic of our survival instinct—and redirect it from physical acquisition to intellectual exploration. Not long ago, adventurers set out to conquer ostensibly savage lands and convert the native populations. Today we reach out with our minds to touch the edge of the universe and the beginning of time. People read about cosmology in much the way that their predecessors read about exotic travels. The big bang serves the same role today that Tahiti or the Congo or Antarctica did a couple centuries ago: It opens the imagination to the exotic magnitude of the world. It feeds the restless mind and spirit with a feeling of adventure, an adventure of a scale and scope unlike anything we've encountered before. Science has always had many of these virtues. Since Einstein's prophecy elevated science to sci/religion and gave it dominion over the universe, however, it has drastically expanded its credibility.

I can even imagine the sci/religious faith expanding to provide a new theory of consciousness that extends this sense of cosmic connection. Cognitive scientists often think of consciousness as an emergent phenomenon, meaning that the brain's processing ability grew greater and greater until it reached a critical threshold at which true consciousness (however we might define it) emerged. The more time I spend with cosmologists, the more I am drawn to a different interpretation. What if consciousness is not something that emerged, but something that exists on a continuum? We all recognize that chimpanzees have less consciousness than humans, cats less than chimps, squirrels less than cats, and so on. To me, it makes philosophical sense that consciousness evolves incrementally out of response to stimulus, all the way down to the simplest bacteria, making it an aspect of life as fundamental as metabolism. But the origin of life, too, had to be in some ways an incremental event. Continuing backward, I think of consciousness rooted in the simplest events and responses, the interactions of subatomic particles and fields. This is a philosophical position, not a testable theory so far as I can tell. But it would give sci/religion something resembling a model of the soul—a sense that our individual consciousness is linked to a universal, eternal responsiveness.

As sci/religion grows ever more generous in scope, old-time religion is struggling to find its place in the new order. But really, this process of readjustment has been going on for centuries. In the fifth century C.E., Saint Augustine already recognized the need to disentangle the biblical account of creation from physical theories of the world, urging that the evidence of the senses should take precedence whenever possible. The Jewish philosopher Moses Maimonides argued that reason should guide our study of the world unless it seemed to contradict the Bible's most fundamental doctrines. Spinoza established the whole framework of cosmic religion in the seventeenth century by identifying God as an unchanging entity that cannot be separated from natural law. In the past century, the Catholic Church has steadily retreated from both cosmology and evolutionary biology. Even biblical literalists who think the world is 6,000 years old feel compelled to use the arguments and evidence of science in a vain attempt to protect the old faith against the onslaught of the new.

All the same, many people remain convinced that science is amoral or even immoral. Large parts of the world are fiercely devoted to old-time religions. As Primack warned, the Church of Einstein might produce the most glorious picture of the universe ever conceived by humans, and yet fail to find its congregation. "How well our cosmology is interpreted in a language meaningful to ordinary people will determine how well its elemental stories are understood, which may in turn affect how positive the consequences for society turn out to be. There is a moral responsibility involved in tampering with the underpinnings of reality," he writes.

Einstein foresaw that traditional religions will have to abandon the idea of a personal God and articulate new moral philosophy. "After religious teachers accomplish the refining process indicated they will surely recognize with joy that true religion has been ennobled and made more profound by scientific knowledge," he wrote. But the success of sci/religion depends even more on its practitioners jettisoning their reticence and speaking openly about the deep mystical satisfaction their work delivers. The material success of science—glitzy consumer electronics, sophisticated new medical treatments—will not make the case for them.

The 1930 manifesto that Einstein wrote for *The New York Times* rings truer than ever today: "Those whose acquaintance with scientific research is derived chiefly from its practical results easily develop a completely false notion of the mentality of the men who, surrounded by a skeptical world, have shown the way to kindred spirits scattered wide through the world and the centuries. Only one who has devoted his life to similar ends can have a vivid realization of what has inspired these men and given them the strength to remain true to their purpose in spite of countless failures. It is cosmic religious feeling that gives a man such strength. A contemporary has said, not unjustly, that in this materialistic age of ours the serious scientific workers are the only profoundly religious people." It is time for the sci/religious faithful to step up to the pulpit and be heard.

LIVERPOOL CITY COUNCIL

ACKNOWLEDGMENTS

I suppose all books are a long journey, but this is my first one; so for me the path has been full of surprise twists, delays, and cathartic moments of clarity. My editor, Stephen Morrow, has done a great job of kicking me through the slow stretches and urging me to trust inspiration whenever it strikes. My coworkers at *Discover* and, before that, at *Scientific American* helped foster many of the themes in this book through innumerable free-form conversations. I am particularly grateful for John Horgan's good-humored arguments, which forced me to think clearly about why he was wrong or even why he might be right. So many friends have lent a sympathetic ear that it would be tedious to thank them all properly. Laurie Shapiro deserves special mention for keeping me focused on the joy of writing. And I owe Michael Abrams for engaging me in a series of sharp debates that, I hope, have steered me away from my more misguided ideas.

Many of the thoughts expressed on these pages have been percolating in my head since childhood. I owe my parents a tremendous debt for their endless intellectual support and encouragement; my mother did a remarkable job of encouraging my youthful interest in astronomy and urging me to keep reading books and attempting to master concepts that seemed beyond my grasp. Over the years my brothers Kevin and Jonathan have helped keep my sense of wonder active with their lively curiosity and keen outsiders' interest in science.

I am amazed at how many researchers have freely given their time and encouragement to help me in my research. Brian Schmidt, Bob Kirshner, Jim Peebles, Andrei Linde, Alan Guth, Mike Turner, Adam Reiss, Robert Jastrow, Neil Turok, Don Nicholson, and above all Saul Perlmutter have been invaluable resources, always available and eager to help out.

Finally, I could not have completed this project without the loyal support of my wife, Lisa Gifford, who married me just as I began to spend my evenings and weekends locked away in front of a computer, lost in a writer's daze. She has been a steady source of comfort and inspiration. Her vivacious spirit lies everywhere between the words of this book.

BIBLIOGRAPHY

Bertotti, B., R. Balbinot, S. Bergia, and A. Messina, eds. *Modern Cosmology in Retrospect.* Cambridge University Press, 1990.

Born, Max. *Einstein's Theory of Relativity.* Dover, 1965.

Calaprice, Alice, ed. *The Expanded Quotable Einstein.* Princeton University Press, 2000.

Christianson, Gale E. *Edwin Hubble, Mariner of the Nebulae.* University of Chicago Press, 1995.

Clark, Ronald W. *Einstein: The Life and Times.* Avon, 1984.

Crowe, Michael J. *Modern Theories of the Universe from Herschel to Hubble.* Dover, 1994.

Danielson, Dennis Richard, ed. *The Book of the Cosmos: Imagining the Universe from Heraclitus to Hawking.* Perseus Publishing, 2000.

Davies, Paul. *God & The New Physics.* Touchstone, 1983.

Dawkins, Richard. *The Blind Watchmaker: Why the Evidence of Evolution Reveals a Universe Without Design.* W. W. Norton & Company, 1996.

Einstein, Albert. *The Principle of Relativity.* Dover, 1952.

———. *Relativity: The Special and General Theory.* Three Rivers Press, 1961.

———. *The World as I See It.* Citadel Press, 1979.

———. *Ideas and Opinions.* Three Rivers Press, 1982.

Ferris, Timothy. *Coming of Age in the Milky Way.* William Morrow & Company, 1988.

———. *The Whole Shebang: A State of the Universe(s) Report.* Touchstone, 1998.

Folsing, Albrecht. *Albert Einstein.* Viking, 1997.

Fox, Everett, trans. *The Five Books of Moses.* Schocken Books, 1995.

Gamow, George. *The Creation of the Universe.* Viking, 1952.

———. *My World Line: An Informal Biography.* Viking, 1970.

Gould, Stephen J. *Rock of Ages: Science and Religion in the Fullness of Life.* Ballantine Books, 2002.

Guth, Alan H. *The Inflationary Universe: The Quest for a New Theory of Cosmic Origins.* Addison Wesley, 1997.

Harper, E., W. C. Parke, and G. D. Anderson. *The George Gamow Symposium.* Astronomical Society of the Pacific, 1997.

Harrison, Edward. *Darkness at Night: A Riddle of the Universe.* Harvard University Press, 1987.

Hawking, Stephen. *A Brief History of Time.* Bantam, 1996.

Hetherington, Noriss S., ed. *Cosmology: Historical, Literary, Philosophical, Religious, and Scientific Perspectives.* Garland Publishing, 1993.

———. *Encyclopedia of Cosmology.* Garland Publishing, 1993.

Hoffmann, Banesh. *Albert Einstein: Creator & Rebel.* Plume, 1972.

Holton, Gerald. *Thematic Origins of Scientific Thought: Kepler to Einstein.* Harvard University Press, 1980.

Hoyle, Fred. *The Nature of the Universe.* Oxford University Press, 1950.

Hubble, Edwin Powell. *The Realm of the Nebulae.* Yale University Press, 1936.

Jammer, Max. *Einstein and Religion.* Princeton University Press, 1999.

Kevles, Daniel. *The Physicists: The History of a Scientific Community in Modern America.* Vantage, 1979.

Kragh, Helge. *Cosmology and Controversy: The Historical Development of Two Theories of the Universe.* Princeton University Press, 1996.

Layzer, David. *Constructing the Universe.* Scientific American Library, 1984.

Lightman, Alan, and Roberta Brawer. *Origins: The Lives and Worlds of Modern Cosmologists.* Harvard University Press, 1990.

Maimonides, Moses. *The Guide for the Perplexed.* Dover, 1956.

Margenau, Henry, and Roy Abraham Varghese. *Cosmos, Bios, Theos: Scientists Reflect on Science, God, and the Origins of the Universe, Life, and Homo Sapiens.* Open Court, 1992.

Munitz, Milton K., ed. *Theories of the Universe: From Babylonian Myth to Modern Science.* Free Press, 1957.

North, J. D. *The Measure of the Universe: A History of Modern Cosmology.* Dover, 1990.

Overbye, Dennis. *Lonely Hearts of the Cosmos: The Story of the Scientific Quest for the Secret of the Universe.* HarperCollins, 1991.

Pais, Abraham. *"Subtle Is the Lord . . .": The Science and the Life of Albert Einstein.* Oxford University Press, 1982.

Perlmutter, Saul, et al. "Cosmology from Type Ia Supernovae," in *Bulletin of the American Astronomical Society* 29: 1351 (1997).

———. "Measurements of Omega and Lambda from 42 High-Redshift Supernovae," in the *Astrophysical Journal,* 517:565 (1999).

Polkinghorne, John. *Faith, Science & Understanding.* Yale University Press, 2000.

Sagan, Carl. *The Demon-Haunted World: Science As a Candle in the Dark.* Ballantine Books, 1997.

———. *Cosmos.* Cosmos Studios, 2000 (DVD and VHS).

Silk, Joseph. *The Big Bang.* W. H. Freeman, 1989.

Spinoza, Benedict de. *Ethics, Including the Improvement of the Understanding.* R. H. M. Elwes, trans. Prometheus, 1989.

Staguhn, Gerhard. *God's Laughter: Physics, Religion, and the Cosmos.* Kodansha, 1994.

Trimble, Virginia. "The 1920 Shapley-Curtis Discussion: Background, Issues, and Aftermath," in *Publications of the Astronomical Society of the Pacific* 107:1133 (1995).

Tropp, E. A., V. Ya. Frenkel, and A. D. Chernin. *Alexander A. Friedmann: The Man Who Made the Universe Expand.* Cambridge University Press, 1993.

Wilson, Edward O. *Consilience: The Unity of Knowledge.* Random House, 1999.

INDEX

A

absolute zero, 152
acceleration, 62
acceleration/deceleration, cosmic, 173, 210, 223–25
Albinoni supernova, 231
Albrecht, Andreas, 202
al-Farghani, 21
alpha-beta-gamma paper, 157
Alpher, Ralph, 156, 157, 158, 169
American Astronomical Society, 241–42, 245
Andromeda nebula (galaxy), 38, 97, 101, 108–109, 110, 117, 118–19, 121
Andromeda supernova of 1885, 161
anthropic principle, 191, 192, 193, 203
antigravity force. *See* Lambda
Aquinas, Thomas (saint), 8, 11, 18, 20
Aristarchus of Samos, 22
Aristotelian model of universe, 19–21, 22, 25
Aristotle, 11, 12, 27, 149
  and empirical data, 18
atom, 54, 55, 127
  primeval, 134, 135, 136, 151, 158, 159, 171
Augustine of Hippo (saint), 8, 11, 24, 31, 40, 44, 77, 90, 207, 257

B

Baade, Walter, 162–63, 213
balloon-borne telescopes, 5, 234
Baronius, Cardinal, 8, 247
Becquerel, Henri, 55
Bessel, Friedrich Wilhelm, 39, 119
Besso, Michele Angelo, 55, 63
Bethe, Hans, 155, 161
Bible (Scripture), 8, 9, 11, 23, 43, 45, 134–35, 158, 168
biblical literalists, 8, 257

big bang cosmology, 8, 18, 32, 44, 136, 144, 158, 160–64, 166, 168, 173, 175–78, 181, 182, 202, 237
  and Christianity, 169–70
"Big Bang Cosmology—Enigmas and Nostrums, The" (Dicke), 190
big crunch, 226
Big Squeeze, 159
Big Throughput Camera, 221
Bondi, Hermann, 164, 165, 172
Book of Genesis. *See* Genesis
Brahe, Tycho, 24, 26
Brief History of Time, A (Hawking), 207, 237, 249
Bunsen, Robert, 42, 108

C

C (creation) term, 167
calculus, 27
Callippus of Cyzicus, 18
Carnegie, Andrew, 115
Carter, Brandon, 191
Casimir, Hendrick, 194
Casimir effect, 194
Catholic Church, 20, 23, 24, 170, 247, 257
Cepheid variable stars, 102–104, 109, 116, 118, 119, 162–63, 212, 244
Cerro Tololo, 221
charge-coupled devices (CCDs), 216–17, 218, 219
Christianity, 18, 20, 169–70
Christianson, Gale, 114
Church of Einstein, 3, 7, 45, 241–58
Clark, Ronald, 49, 69
Clerke, Agnes, 98
Coma cluster, 187
Commentariolus (Copernicus), 22
computer, electronic, 160, 169, 216, 217, 219
"computers" of stellar astronomy, 102
"Concerning the Investigation of the State of Ether in the Magnetic Field" (Einstein), 51
Confessions (Augustine), 207
Consciousness, 256
conservation laws, 41, 166, 167
Copernican system, 25, 27, 28
Copernicus, Nicolaus, 20, 45, 74
  heliocentric model of, 22–23
cosmic rays, 136, 151
"Cosmological considerations of the General Theory of Relativity" (Einstein), 69, 75, 79, 138
cosmological constant. *See* Lambda.
cosmological principle, 74, 128–29, 164, 185
cosmological redshift. *See* redshift
cosmologists, 4–5, 6, 7, 168
  theoretical, 6, 232, 233

cosmology, 136, 145–146, 150, 165, 194
  and God, 9
  Soviets and, 171–72
  spherical shell, 16–18
creation, spontaneous, 165, 166, 167
creationists, beliefs of, 9, 10
Critique of Pure Wisdom (Kant), 50
Curtis, Heber, 98–99, 100–101, 104

D

dark energy, 6, 227, 231, 234
dark matter, 5, 186–90, 192
dark sky question, 34–35, 137
Darwin, Charles, 79
Dawkins, Richard, 250, 253
De Revolutionibus (Copernicus), 22–23, 25
de Sitter, Willem, 67, 82–83
  on general relativity, 83–84, 86
  on Lambda, 86
  on universe, 83–84, 122, 123, 141
de Sitter effect, 85, 95, 127
de Sitter universe, 83–84, 85
deceleration parameter, 210, 223
Democritus, 55
Descartes, Rene, 77, 85, 129
design, intelligent, 10
deuterium, 178
dialectical materialism, 154
Dialogue (Galileo), 25
Dicke, Robert, 159, 175, 176, 183, 190
Dingle, Herbert, 149
Doppler, Christian Johann, 107
Doppler shift, 107–108, 142
Duncan, John, 118
Dust, 230, 231
Dyson, Frank, Sir, 67

E

$E=mc^2$, 59, 167, 193, 204, 226
Earth
  age of, 40–41, 131
  geological history of, 8
  position of in Milky Way, 101–102, 103–104
Eddington, Arthur Stanley, 67, 68, 85, 95, 132–36, 139, 151–52, 159
Ehrenfest, Paul, 87
Einstein, Albert, 15, 18, 20, 21, 29, 32, 45, 111, 249, 254
  on acceleration, 62
  arrogant detachment of, 78
  celebrity of, 92–93
  on cosmology, 48, 69–80,140–42, 171
  deism of, 50
  on energy, 58–59
  on expanding universe, 140–41, 142, 181
  and flat universe, 92, 141, 183
  on Friedmann, 92, 93

Einstein, Albert (cont.)
and general theory of relativity, 48, 61, 64, 66, 138
on gravity, 48, 49, 61–62, 64, 137, 138
and Lambda (cosmological constant), 5, 7, 13, 48–49
life of, 47, 49–51, 82, 139
on light, 51, 53, 54, 56, 57–58, 59, 61
and origin of universe, 142, 143
overreaching of, 78–79
on physics, 56
published papers of, 56
on quantum theory, 137–38
on redshifts, 139, 140, 142
and relativity, 7, 12, 13, 51, 56, 57, 58, 59, 138
and sci/religion, 3, 13–14, 43, 135, 258
on space, 48, 64–65, 72
and special theory of relativity, 47, 56, 58, 60
and unified field theory, 129, 138, 197
on universe, 69–75, 181
Einstein, Hermann, 49
Einstein, Pauline (née Koch), 49
Einstein-de Sitter universe, 92, 141, 142, 144, 171
electromagnetism, 137, 138, 198
electron, 55, 59
energy and matter, 58–60
epicycles, 21, 23
ether, 19, 20, 24, 27, 42, 50, 235
luminiferous, 53, 59
ethics and morals, 11
Eudoxus of Cnidos, 16, 45, 185
evolution, 8–9, 254

F

faith, 10, 11, 44, 45, 76, 77, 79, 80, 135, 250, 253
falsification (falsifiability), 170, 177, 198, 255
Fermi, Enrico, 136
Feynman, Richard, 193
Findlay-Freundlich, Erwin, 62
"Foundation of the General Theory of Relativity, The" (Einstein), 64
flatness problem, 182, 200
Freud, Sigmund, 79
Friedmann, Alexander, 82, 86–88, 91, 93–94, 237
on age of universe, 90
on oscillating universe, 158–59
on relativity, 88–89, 92
Friedmann-Einstein model of universe, 141

G

galaxies, 38, 77, 97, 98, 100, 108, 142, 173–74, 187, 188, 212, 213, 215
distance of, 124–25, 213
formation of, 165, 168, 178, 201

spiral nebulae as, 119, 120, 121
spiral, 124, 188
Galileo Galilei, 8, 18, 20, 43, 45, 247
observations of, 24–25
Gamow, George, 143, 153–54, 155–60, 168, 170, 171, 175
general theory of relativity, 7, 12, 29, 34, 48, 49, 54, 61, 64, 66, 75, 138, 142, 146, 181
de Sitter on, 83
test for, 62, 65, 66, 67–68
Genesis, 31, 40, 134, 158
geocentric cosmology, 23
Glashow, Sheldon, 198
God, 45, 191, 239
and big bang, 169
concepts of, 9, 10, 11, 12, 13, 77, 197, 207, 252
Einstein and, 13, 43–44, 45, 48–51, 63, 69, 79–80, 86, 120, 138, 171, 183, 185, 252
gravity and, 28
Greek cosmology and, 21
Hawking on, 191, 193, 249
Lemaître on, 135
Linde on, 205, 248–49
and mathematics, 19–20, 252
Milne on, 149, 169
Newton and, 29, 30, 31, 32, 33, 34, 43, 44, 48, 58
Perlmutter on, 249–50
sci/religion and, 248–49
God and the Astronomers (Jastrow), 9
gods, of Greeks, 15–16
Gold, Thomas, 164–65, 172
Gould, Stephen Jay, 248
grand unified theory, 198
gravitational drag, 147
gravity, 12, 13, 28, 29, 30–33, 36, 43, 45, 47, 52, 53, 128, 137
and effect on light, 61–62, 65, 67–68, 72
Einstein on, 48, 49, 61–62, 64, 137, 138, 195
Great Debate, 98–111
Greek philosophers, 15–22, 58
Griest, Kim, 190
Grossman, Marcel, 61
Gunn, James, 125
Guth, Alan, 32, 92, 189, 198–200, 202, 204–208, 226, 232, 234, 241, 245

H

Hale telescope, 212
Halley, Edmond, 27, 34
Hawking, Stephen, 78, 90, 126, 191, 192, 201, 207, 237, 249
heavy elements, creation of, 155, 156, 160, 161, 168, 211
Heisenberg, Werner, 137
heliocentric model of universe, 22–23, 25
responses to, 24

Helmholtz, Hermann von, 41–42
Herman, Robert, 157, 169
Herschel, William, 37–39, 41, 77, 97, 109, 117
Hertzsprung, Ejnar, 102, 109
High-Z Supernova Search, 211, 218, 222, 225, 227, 228
Hooker, John, 115
Hooker telescope, 115–16, 124, 244
horizon problem, 186, 200–201
Hoyle, Fred, 160–61, 163, 164–69, 171, 172, 174, 178, 196, 213
Hubble, Edwin, 95, 111, 139–40, 141, 147–50, 152, 162, 212
  desciption of, 113–15
  early life of, 115–16
  and galaxy distances, 124–26
  and nebulae, 115, 117–19
  as nontheorist, 126–127
Hubble, Grace, 114
Hubble constant, 214–15
Hubble's law, 125, 128
Huggins, William, Sir, 42–43
Humason, Milton, 124, 128, 140, 173, 213
hydrogen, 154–55, 178

I

inertia, 12, 13, 24, 70, 84, 238
infinity, 32, 33, 35
inflation, 32, 199, 200–204, 206, 207, 232–36
  chaotic, 205
intelligent design, 10
Internet, 217
"Is Science a Religion?" (Dawkins), 251
"Is the Universe a Vacuum Fluctuation?" (Tryon), 197
island universe, 71, 100, 104, 109, 117

J

Jastrow, Robert, 9
Jeans, James, 147

K

Kant, Immanuel, 36, 50
  on Milky Way, 36–37, 77
Keck Observatory, 2, 4, 221, 244
Kelvin, Lord, 41
Kepler, Johannes, 17–18, 45
  planetary system of, 26–27
Kepler's Supernova, 161, 215
kinematic relativity, 148, 149
Kirchhoff, Gustav Robert, 42, 108
Kirshner, Robert, 211, 218, 223, 225, 228, 229, 231, 232, 236, 237, 247–48
Kragh, Helge, 131
Krutkov, Yuri, 93

L

Lambda (cosmological constant), 5, 6, 7, 13, 20–21, 32, 33, 44, 75, 83, 88, 92, 105, 133, 137, 143–44, 148, 167, 179

Einstein and, 48–49, 75, 76, 78, 80, 84, 86, 93, 114, 128, 141, 142, 143, 237
and sci/religion, 8, 14, 143
Lambda, rehabilitation of, 193, 195–96, 199, 202, 207–208, 223, 224, 226, 231, 233–35
Leavitt, Henrietta Swan, 102-103
Lemaître, Georges, 82, 94–95, 159, 170, 196
on expanding universe, 130–32, 137
on God, 135
on origin of universe, 133–34, 135, 136, 151, 152
on relativity, 95–96
Leucippus, 55
Leverrier, Urbain, 65
light
gravity's effect on, 61–62, 65, 67–68, 72
propagation of, 51, 52–53
speed of, 53, 57–58, 59–60, 83
light curve, 220, 221, 225
light waves, 107, 147
Linde, Andrei, 201–203, 205 248–49
Lowell Observatory, 106
Lowell, Percival, 106–107
Lowenthal, Elsa, 82
luminosity, 102, 103, 118, 212, 213, 220
Lundmark, Knut, 118
Luther, Martin, 24, 113

M

Mach, Ernst, 54, 60, 127
Magellanic Clouds, 110
magnetic monopoles, 198, 199
Maimonides, Moses, 11, 257
"Manifesto to Europeans," 64
"Manifesto to the Civilized World," 63
Maric, Mileva, 55
Mars, canals of, 106
Mason, Brian, 242, 245–46
Mathematical Principles of Natural Philosophy (Principia) (Newton), 28
Mauna Kea, 1–2
Maxwell, James Clerk, 54
Mayer, Walther, 138
McCrea, William, 167
Mercury, orbit of, 52, 65, 66
Metaphysica (Aristotle), 20
Michelson, A. A., 51, 53
microwave background, 157, 175–78, 181, 184, 200, 233–34, 237, 245
Milky Way, 35, 38, 74, 96–98, 100, 101, 103, 104, 109
Kant on structure of, 36–37, 77
Milne, Edward A., 147–49, 168
Minkowski, Hermann, 60
Minkowski, Rudolph, 213
Misner, Charles, 184, 185
"Modern Aristotelianism" (Dingle), 149
moral relativism, 58

morals. *See* ethics and morals
Morley, Edward, 53
Moulton, Forest Ray, 115
Mount Palomar, 163, 212
Mount Wilson observatory, 114, 115, 139, 244
My World Line (Gamow), 143
Mysterium Cosmographicum (Kepler), 26

N

natural law, uniformity of, 42, 43
Nature of the Universe, The (Hoyle), 169
nebulae, 37, 38, 40, 42, 77, 80, 84, 95, 97–100, 104–108, 114, 123
  classification of, 117–18
nebulae, "planetary," 117
nebulae, spiral, 115, 117, 124
  as galaxies, 119, 120
  motion of, 107, 108, 110, 122
  speed of, 108, 109, 110, 122
negative (potential) energy, 204
Neptune, 65
Nernst, Walter, 193
neutrinos, 190
neutrons, 151, 155, 156, 178
Newcomb, Simon, 66
Newton, Isaac, 12, 27, 45, 59, 90
  on gravity, 28–29, 30, 31, 34, 48, 52, 64
  on infinity, 32, 33
  and Sensorium of God, 54
  on universe, 30, 31, 32, 33, 38, 43, 47
no-boundary proposal, 207, 249
"Note on de Sitter's Universe" (Lemaître), 95
nuclear forces, 138, 198
nuclear physics, 156, 168
nuclear theory, 151, 156

O

observation, 10, 18, 21, 25, 26, 37, 63, 66–68, 76, 97, 127, 166, 172, 198, 212, 220
observatories, satellite, 237
Olbers, Heinrich Wilhelm, 34
Olber's paradox, 34–35, 52, 70, 96, 137
"On the Curvature of Space" (Friedmann), 89
"On the Electrodynamics of Moving Bodies" (Einstein), 56
On the Heavens (De Caelo) (Aristotle), 19, 20
"On the Possibility of a World with Constant Negative Curvature" (Friedmann), 89
optical interferometry, 244
Optics (Newton), 59
Ostriker, Jeremiah, 188

P

Pais, Abraham, 68
parallax, 39, 40
Parsons, William, 97
particle physics, 136
particles, virtual, 194–96, 233

particles and waves, 59
Paul V (pope), 24
Pauli, Wolfgang, 193
Payne-Gaposchkin, Cecilia, 68, 119, 154
Peebles, James, 175–78, 183, 188, 203, 228
Pennypacker, Carl, 211, 215
Penzias, Arno, 175–77, 181
Perlmutter, Saul, 209–11, 214–19, 222–25, 227, 230, 231, 235, 246
  and expansion of universe, 4–5
  on God, 249–50
  and Lambda, 5–6
Pius XII (pope), 169–70, 171
Planck, Max, 60, 63
Plato, 17
Pluto, 106
Primack, Joel, 251, 257
primeval atom, 134–36, 151, 158, 159, 171
Principia (Newton), 28, 29
protons, 155, 178
Proxima Centauri, 22
psi, 3
Ptolemaeus, Claudius (Ptolemy), 21
Pythagoras, 17

Q

quantum physics, 51, 56, 59, 130, 135, 137, 151, 155, 193
quantum theory, 3, 137, 234
quintessence, 235, 236

R

radiation, background, from big bang, 157, 175–77, 181, 185, 186, 200, 233–34, 237, 245
radiation pressure, 130, 132, 151
radio astronomy, 173–74
radioactive decay, 55, 131
Realm of the Nebulae, The (Hubble), 128
redshift, 122-27, 146–47, 150, 178, 211–13, 219
  cosmological, 142
"Relation Between Distance and Radial Velocity Among Extra-Galactic Nebulae, A" (Hubble), 126
relativism, moral, 58
relativity
  critiques of, 82
  general theory of, 7, 12, 29, 48, 49, 61, 64–66, 75, 83, 138, 142, 146, 181
  kinematic, 148, 149, 171
  special theory of, 47, 56, 58, 60
Relativity (Einstein), 72
religion
  and empirical study of world, 8, 11
  and science, 169–70, 249, 250–51
religion, old-time, 2, 13, 79, 170, 253
  retreat of, 11

religion/science, demarcation line for, 8, 247
Riemann, Bernhard, 71
Riess, Adam, 220, 225, 230, 231, 239
Robertson, Howard, 164
Roentgen, Wilhelm, 55
Rubin, Vera, 188
Russell, Bertrand, 18
Russell, Henry Norris, 121
Ryle, Martin, 174

S

Sagan, Carl, 250
Salam, Abdus, 198
Sandage, Allan, 163, 173, 212, 214
satellite observatories, 237
saving the appearances, 16–17
scalar field, 235, 236
Schmidt, Brian, 5, 210, 211, 218, 219, 222, 225, 228, 230, 235, 238, 250
Schwinger, Julian, 193
sci/religion, 4, 6, 25, 33, 63, 76, 79, 81, 98, 170
  and aggression, 255
  and awe, 251
  and consciousness, 256
  description of, 2–3, 255
  and ecstasy, 12, 254
  and faith, 10, 80
  and God, 248–49
  and hope, 253
  language of, 242–43, 244
  and morality, 254, 257
  science transformed into, 7
  temples (observatories) of, 243, 244–45
science, religion and, 169–70, 249, 250–51
Science (magazine), 6, 225
science/religion, demarcation line for, 8, 247
Scripture. *See* Bible
Sensorium of God, 54, 57
Shapley, Harlow, 95, 98–99, 102, 103–104, 105, 116–17, 119, 121
Silberstein, Ludwik, 123
61 Cygni, 39, 119
Slipher, Vesto Melvin, 106–110, 117, 123
Smolin, Lee, 206
solar spectrum, 42
solution B, 83–84, 86, 122, 130, 199
sound waves, 107
Soviet Union, cosmology in 171–72, 194, 202
space
  absolute, 43, 54, 57, 58, 57, 70
  curved, 48, 64–65, 72, 75–76, 89, 141, 148
  Einstein on, 48, 64–65
  models of, 90
space and matter, interaction of, 48, 49
space and time, 54, 58, 71
space-time, 7, 12, 60, 84

special theory of relativity, 47, 56, 58, 60
spectrum, 42, 108
speed of light, 53, 57–58, 59–60, 83
spherical shell cosmology, 16–18
Spinoza, Baruch, 11, 77, 251–52, 257
Starobinsky, Alexei, 202
stars
  composition of, 42, 154–55
  nuclear reactions in, 155
steady state cosmology, 165–69, 171, 172, 173, 174, 177, 201
Steinhardt, Paul, 202, 235, 236, 237
sun
  age of, 41
  composition of, 42
Supernova Cosmology Project, 211, 215, 218, 221–25, 227, 228
supernovas (exploding stars), 4, 6, 101, 161–62, 168, 211, 213, 215, 216, 219, 220, 223, 226, 228, 229
  distant, 218, 230–31
  searching for, 220–22
  types of, 213–14

T

Tammann, Gustav, 214
technology, 216–217
telescopes
  balloon-borne, 5, 234
  Hale, 212
  Keck, 2, 4, 221, 244
Teller, Edward, 136
theologians, 8, 11
Thomson, J. J., 55
Tolman, Richard, 150, 152–53
Tombaugh, Clyde, 106
Triangulum nebula, 118, 119
Tryon, Edward, 196–97
Turner, Michael, 192, 227, 233
Tycho's Supernova, 161
Type I supernovas, 213, 214
Type Ia supernovas, 214, 215, 219–21, 228–31
Type II supernovas, 213–14

U

uncertainty principle, 137
unified field theory, 129, 139
unified physics, 198
uniformity
  cosmic, 184–85, 186
  of matter, 129
  of natural (physical) law, 42, 43, 121–22
universal gravitation. *See* gravity
Universal Natural History and Theory of the Heavens (Kant), 36
universe
  accelerating, 224, 225, 228, 238–239
  age of, 90–91, 131, 143, 162–63, 178, 181, 227

universe (cont.)
Aristotelian, 19–21, 22, 25
of de Sitter, 83, 84, 85, 122, 127, 130
density of, 74, 83, 182, 183, 189, 200, 202–203, 223–24, 226, 227, 232, 233, 235
dynamic, 89–93, 96
of Einstein-de Sitter, 92, 141, 142
Einstein on, 69–75, 140
evolution of, 153, 158, 159, 163, 184
expanding, 34, 35, 85, 91, 114, 122, 125–26, 128, 130–33, 140, 142–44, 146, 148, 163, 165, 181
expansion of, speed (rate) of, 4–5, 6, 131, 210, 213, 223, 224, 225
as finite, 12, 13, 19, 43, 48, 71, 75, 77
as finite in age, 137
as finite in size and age, 35, 137
flat, 91–92, 141, 144, 182–83, 208, 234
of Friedmann-Einstein, 141
heliocentric, 22–23
as infinite, 12, 24, 29, 30, 31, 32, 33, 37, 38, 40, 48, 70
of Lemaître, 130
mass of, 226
origin of, 132, 133–34, 142, 143, 151, 158
oscillating (phoenix), 89, 90, 141, 158–59, 205
past state of, 150, 155, 156, 183, 230, 233, 246
size of, 21, 24, 74–75, 119, 137
spherical shell model of, 16–18
steady state, 165–69, 171–74, 177, 201
thermodynamics of, 152
as unchanging, 31–33, 75, 77, 78, 111, 122
"Universe, Life, Consciousness" (Linde), 248
universes, multiple (multiverse), 192, 205–206
Uranus, 38, 66, 119
Urban VIII (pope), 25
Ussher, James (archbishop), 41

V

vacuum energy, 193, 195–98, 207–208, 224, 227, 232, 234, 236
van Maanen, Adriaan, 120
variable stars, 101, 102, 103, 118
virtual particles, 194, 195, 196, 233

W

wavelength, 107–108
Weinberg, Steven, 192, 198
Westfall, Richard S., 27
Wheeler, John A., 191
white dwarf, 214, 229
Wilson, Robert, 175–76, 181
WIMPS (weakly interacting massive particles), 189

Wirtz, Carl, 123
World as Space and Time, The (Friedmann), 90
Wright, Thomas, 31, 36

X

X rays, 55

Y

ylem, 156

Z

Z (redshift), 211
Zeldovich, Yakov, 194–95, 196
Zwicky, Fritz, 136, 146–47, 187